T0362724

the Australian Beekeeping Manual

3rd edition

Thanks to Erica Siegel, wildlife photographer and native fauna enthusiast, for writing the section on native solitary bees in Chapter 14: Native Bees.

First published 2015
2nd edition published 2020
This edition published 2023

Exisle Publishing Pty Ltd
PO Box 864, Chatswood, NSW 2057, Australia
226 High Street, Dunedin, 9016, New Zealand
www.exislepublishing.com

A CiP record for this book is available from the National Library of Australia.

ISBN 978 1 922539 79 3

Designed by Tracey Gibbs
Additional typesetting by Shaun Jury
Typeset in Beaufort 10.5pt / 15pt
Printed in China

This book uses paper sourced under ISO 14001 guidelines from well-managed forests and other controlled sources.

10 9 8 7 6 5 4 3

Disclaimer
While this book is intended as a general information resource and all care has been taken in compiling the contents, neither the author nor the publisher and their distributors can be held responsible for any loss, claim or action that may arise from reliance on the information contained in this book.

ALSO BY EXISLE PUBLISHING

PRACTICAL BEEKEEPING IN NEW ZEALAND (5TH EDITION)

Andrew Matheson
& Murray Reid
ISBN 978 1 77559 362 1

MANUKA: THE BIOGRAPHY OF AN EXTRAORDINARY HONEY

Cliff Van Eaton
ISBN 978 1 77559 163 4

the Australian Beekeeping Manual

3rd edition

ROBERT OWEN

EXISLE PUBLISHING

Contents

Introduction

It has been eight years since the first edition of this book was published. During this time many Australian and overseas beekeepers have written to say how useful and informative the book has been. The first edition quickly became the most widely purchased book on beekeeping in Australia, which shows its usefulness to both hobby and professional beekeepers. During the last five years, Flow Hive has become a major success both in Australia and globally, contributing to the increase in popularity of this fascinating hobby. The second edition of this book therefore added a chapter on Flow Hive, aiming to give beginner beekeepers an overview of the hive. In June 2022, the parasitic mite *Varroa destructor* was, unfortunately, discovered in Newcastle, New South Wales. *Varroa destructor* is the biggest health challenge facing beekeepers globally, and so the third edition of this book includes additional information on the management of this pest.

The majority of beekeeping books currently for sale are either published in the United States or in the United Kingdom where beekeeping conditions are very different to those experienced in Australia. While there have been some excellent books previously published in Australia, I felt the time was right for an alternative set of insights and experiences which I hoped would be of interest to the beginner and the more experienced beekeeper. During the many years I have been a beekeeper, I have had the opportunity to speak with numerous beekeepers about their varied beekeeping experiences as well as listening and responding to their practical difficulties. This book is a culmination of those varied discussions. It is my hope that the third edition of this

0.2

0.1: Drinking honey.

0.2: Bee on a climbing rose.

book will continue to assist beekeepers and will answer many of the questions they may have.

A keen beekeeper will work to obtain information from many sources on how best to keep bees. These may include books, the Internet, YouTube, local beekeeping clubs and talking with other beekeepers. Glance through a number of beekeeping books or Internet sites and you will notice that beekeeping practices around the world differ. But despite these differences, many of the practices and tools used are similar globally. This reflects a common beekeeping heritage that started in the United States and Europe. Between 1851 and 1865 there were five inventions that changed traditional beekeeping management and practices from an inefficient backyard hobby to the more efficient practices that are used today. These inventions are:

Bee space – commonly attributed to Reverend Lorenzo Lorraine Langstroth (the father of American beekeeping) in 1851.

The Langstroth hive – patented in 1852 by Reverend Lorenzo Lorraine Langstroth, although a similar hive design, the bee space, and removable frames had been invented in Europe previously.

The smoker – invented by Moses Quinby in 1853 in the United States.

Comb foundation – invented by the German beekeeper J. Mehring in 1857.

The centrifugal extractor – invented in 1865 by Major D. Hruschka of Venice, Italy.

These inventions made it possible for the first time to efficiently manage bees and to remove frames from a hive either to inspect or to extract honey. Previous hives were often completely destroyed to obtain the honey collected by the occupying colony of bees. Between 1865 and 2015 little changed in beekeeping practice except perhaps migratory beekeeping and the management of disease. In 2015, with the introduction of Flow Hive, an Australian invention, beekeepers needed to rethink their management practices and consider alternative ways of managing bees. This has been good for the industry, which, until this time, had retained its traditional view of best practices. Whichever way your knowledge has been gained, I would strongly recommend you look at all techniques used globally and select the ones you believe are best for your own and your bees' situation. Do not become dogmatic about techniques! In the early stages of beekeeping you will be bombarded with differing views on management but try to absorb the information and, as your knowledge and experience grow, you can adopt or discard the information that suits you and your own hives.

This book, as well as describing the more traditional methods of keeping bees in Langstroth hives, also includes information on other types of hives including Flow Hive and natural beekeeping techniques that are becoming increasingly popular among hobby beekeepers. Natural beekeepers usually have strong philosophical opinions as to why their techniques are best for the bees in their care. There are many advantages and some disadvantages of natural beekeeping methods, both for bees and the beekeeper. The chapter on natural beekeeping will, I hope, provide useful information for those wishing to join this growing movement. Similarly, with Flow Hive, many beekeepers are using this hive and many of the hive-management techniques are different to those for Langstroth hives. Chapter 20 in the book is intended to provide useful information on the management of bees and honey harvesting using Flow Hive.

Beekeeping is a rewarding and addictive hobby. I have not yet met a beekeeper who does not find a hive and the activities of the bees that call it home, with all its variety and complexity, to be a topic of fascination. It's a good topic of discussion with other beekeepers and friends, as well as a source of continual learning. Recently, beekeeping has become increasing popular. This is due partly to increased publicity both in Australia and overseas of the increasing problems bees are facing due to environmental challenges and beekeeping practices. Another aspect is the trend for more sustainable and lower environmental impact living that is increasingly gaining a foothold. Many people are developing an interest in growing their own food or in sourcing organic produce untouched by harmful chemicals and other artificial elements. The increased knowledge of food plants and their pollination requirements usually follows with some recognition of the value the amazing honey bee adds to our lives. Not only does the bee produce the honey we so love to eat but it also has unparalleled supremacy as a pollinator of the fruits and vegetables we enjoy on a daily basis. With a growing realisation of the importance of the bee it is a small step for someone to then become interested in keeping their own hives.

For me, even updating the third edition of this book has been a learning experience as I set out to try for myself most of the equipment and techniques explained in the following pages. I found that much of what other beekeepers had told me worked as they had explained, although equally some techniques did not work very well and I have not recommended their use in this book. Many common overseas practices that are not used in Australia also worked well and should be adopted by more beekeepers here. I have recommended these in the relevant parts of the book.

Some may feel information presented in the following pages is overly influenced by my own experiences of beekeeping in the south-eastern state of Victoria. Rest assured that the management and practical skills I describe apply equally well throughout the rest of the country, but the timing may need some modification due to local climatic conditions.

Finally, this book would not have been possible without the advice and input of many of my fellow beekeepers. The list of people who have generously contributed both time and information is provided at the end of the book. Any errors that have been made are my responsibility either for not fully understanding a topic or for not listening closely enough when techniques and practices were explained to me by those more knowledgeable on that subject than I. There is no 'one' or 'right' way of keeping bees and your way of doing things may differ to mine. Differences among beekeepers should be viewed as an opportunity to expand and develop knowledge and skills. Our hobby is fortunately one where knowledge is never complete and the bees will continue to teach us until we light our last smoker and pick up our last hive tool.

If any errors are noticed or you have suggestions for improvement, please send them to:
BeeDiseases@gmail.com

Robert Owen
Faculty of Veterinary and Agricultural Sciences
University of Melbourne

1.1

1.2

1.3

Life history of the honey bee

There are many books available on the life history, biology and anatomy of the honey bee. The web is also a mine of information for beekeepers wishing to expand their knowledge of this subject. The aim of this chapter is not to provide a detailed biology or life history of the honey bee, but to provide some basic information that can be used to better understand the content of this book. For more detailed information the reader is referred to the many excellent resources listed in the bibliography.

CASTES OF THE HONEY BEE

In each bee colony there are three castes of bee.

A SINGLE FEMALE QUEEN

There is usually only a single queen in each hive or colony and one of her main roles is to lay eggs. Once an egg has been laid by the queen she has no other role in the upbringing of the brood. The female nursery workers take over all responsibility for feeding and care of the young bee. Queen bees are larger than either female workers or male drones. Queens typically live between one and three years. Contrary to common belief, queens take no part in the decision making of the colony and all major decisions, such as how many drone or worker eggs are laid, or whether to swarm, are actually made collectively by the worker bees.

1.1: A Caucasian bee, *Apis mellifera caucasica*. Notice the longer, slender abdomen of the queen compared to the female worker bees next to her. This allows the queen to deposit eggs at the bottom of empty brood cells.

1.2: A female worker bee.

1.3: Although this photo appears to show a bee covered in pollen, it is in fact a fly of the order Diptera. The antennae of a fly are characterised by an arista, a hair like structure which the photo shows coming off one of the antennal segments. The arista is used to detect changes in temperature and humidity.

1.4: A male drone bee. Notice the larger eyes and the shorter, stubbier abdomen compared to the female worker bees next to him.

1.4

TENS OF THOUSANDS OF FEMALE WORKERS

A typical colony consists of at least 95 per cent workers and depending on the time of year between 20,000 and 60,000 workers may be present in a hive. Physically, workers are the smallest caste of honey bee. They also have a very limited ability to lay small numbers of unfertilised drone eggs if the colony becomes queenless. Workers perform all of the work in the hive, including rearing brood, cleaning cells, guarding the hive entrance and foraging. During the summer a typical worker lives between 30 and 50 days after emerging from her cell although this period is considerably longer in very cold areas where the bees are unable to leave the colony.

UP TO ABOUT 1000 MALE DRONES

The only known role of a drone is to mate with queens from other colonies. Drones are usually intermediate in size between workers and mature queens and are easily differentiated from workers by their larger eyes and shorter, stubbier abdomen. Drones live about six weeks after emerging from their cells and are usually only found in colonies during the spring and summer.

It does not take long to be able to tell the difference between female workers and male drones, but queens are usually more difficult to find. They are the largest bee in the colony and can be recognised by their long, tapered abdomen which is used to deposit eggs at the bottom of empty brood cells. The main difficulty faced by the beekeeper when attempting to find a queen is that there is only one of them hiding among tens of thousands of other bees.

1.5

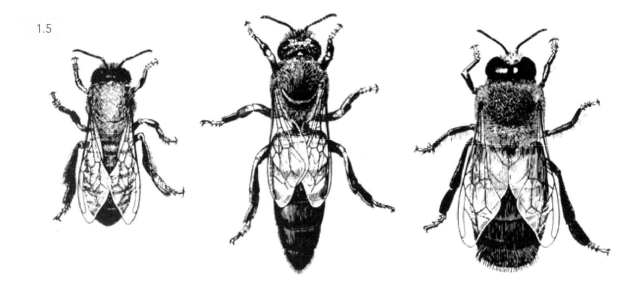

1.5: A female worker, queen, male drone — three castes of honey bees.

THE WESTERN HONEY BEE AND ITS SUBSPECIES

First, a note about the difference between the names Western honey bee and European honey bee. The Western honey bee includes all of the races (or subspecies) of *Apis mellifera* and is found in Africa, Europe, the Middle East and regions much further east. European honey bees are the races of the Western honey bee, *Apis mellifera*, found in Europe and parts of the Middle East. In Australia, the name European honey bee is used to denote the Italian, Caucasian, Carniolan and dark German races of bees.

The Western honey bee, *Apis mellifera*, is one of twelve currently recognised species of honey bee found throughout the world. In spite of this it produces most of the commercial honey consumed globally. The reasons for its success are twofold. Firstly, after thousands of years of human harvesting and selection the European races of *Apis mellifera* have been spread around the world and selectively bred for their docility and high honey yield. Secondly, the other eleven honey bee species are geographically isolated in parts of Asia and the Middle East. All of these eleven species have their ancestral home in South East Asia or India.

The European honey bee was first introduced into Australia in 1822. Since then it has spread to most parts of the country except for the more inhospitable desert regions. The European honey bee races found in Australia today are races or subspecies of *Apis mellifera*. The ancestors of the European honey bee spent most of the last 70,000 years of their existence living in their homelands of Europe, middle Europe, Eastern Europe and the foothills of the Ural Mountains near the Caspian Sea.

1.6

1.7

1.8

1.6: The Italian bee, *Apis mellifera ligustica*, is more yellow or straw-coloured than either Caucasians or Carniolan bees. Notice the dark Caucasian bee to the right of the queen that has drifted in from another hive.

1.7: The Caucasian bee, *Apis mellifera caucasica*, from the Caucasus Mountains region of the former Soviet Union.

1.8: The Carniolan bee, *Apis mellifera carnica*, from the region around Austria and the former Yugoslavia in south-eastern Europe.

The traditional races of the European honey bee found in Australia are:

- *Apis mellifera ligustica* – Italian bees
- *Apis mellifera caucasica* – Caucasians
- *Apis mellifera carnica* – Carniolans
- *Apis mellifera mellifera* – often inaccurately called the dark German bee. It is in fact native to large areas of Europe from south-east France and Spain to Scandinavia. Although the European honey bees found in Australia are usually a mixture of all four subspecies of bees, feral bees are predominantly *Apis mellifera mellifera* in their genetic origins.

The characteristics of these races of bees that are of interest to beekeepers are detailed in Chapter 5: Your First Bees.

The four races of *Apis mellifera* found in Australia can interbreed with each other in order to produce viable offspring. Since all four races were introduced into Australia at different times since 1822 there has been extensive interbreeding among them. While the original four races of bee often have a different colour and possess different

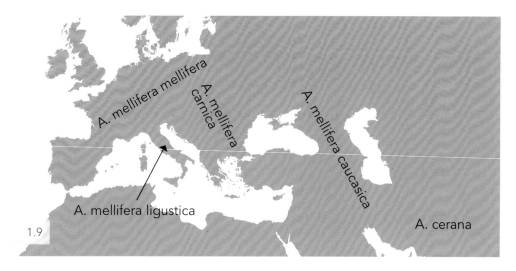

1.9: Geographic origins of the European honey bee.

characteristics, the Australian honey bee is a mongrel mix of different races. In South Australia, Kangaroo Island beekeepers claim the purest Italian Ligurian bees globally. Research performed at the University of Sydney shows that Kangaroo Island bees are the purest Ligurian bees in Australia. In contrast, the highlands of Tasmania are home to the most pure *A. m. mellifera* in Australia.

LIFECYCLE OF THE HONEY BEE

For a colony to survive there needs to be a healthy queen laying eggs that develop into adult worker and drone bees. Queens do not live forever and a new queen is also required when the colony swarms, taking with them the existing queen.

There are three situations in which a new queen is required — swarming, supersedure and emergency.

SWARMING

PREPARING FOR SWARMING

Triggered by the increased flow of nectar during the spring and also by the lengthening hours of daylight, the queen starts to lay many more eggs, between 1000 and 2000 eggs per day, causing the number of bees in the colony to increase rapidly. Soon the hive becomes overcrowded and congestion is one of the major triggers for the colony to swarm (but not the only trigger). Although the first signs of swarming noticed by most beekeepers are queen cells appearing on brood frames, a colony's preparations for swarming actually start around a month before this. About three weeks before the colony starts building queen cells, the workers begin constructing drone cells into

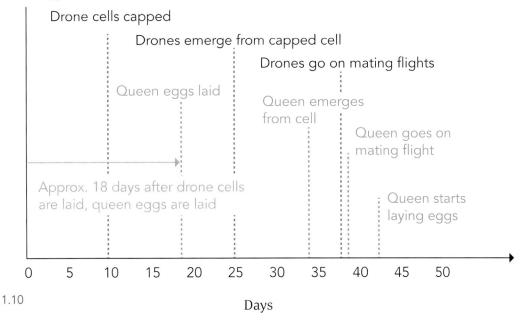

Drone eggs laid

Drone cells capped

Drones emerge from capped cell

Drones go on mating flights

Queen eggs laid

Queen emerges from cell

Queen goes on mating flight

Approx. 18 days after drone cells are laid, queen eggs are laid

Queen starts laying eggs

0 5 10 15 20 25 30 35 40 45 50

1.10

Days

1.11

1.10: The days that the drones go on their first mating flights and the queen goes on her mating flight are about the same. This means the colony starts preparing drone cells about 18 days before the first queen cells can be seen. Once you first see drone cells you know that the colony will start rearing queens in the next two to three weeks.

1.11: Drone cells in a hive are a sign that the colony is planning to swarm, even before the first queen cells can be seen.

which male drone eggs are laid by the old queen. Drone cells are larger than their worker cell counterparts as they are required to hold the much bigger drone pupae and are typically located on the lower half of brood frames.

A few weeks after the queen has started to lay more drone eggs the workers build about six to eight queen cells, usually at the bottom of the brood frames. Normal worker, drone or honey cells are almost horizontal on the frame; queen cells hang vertically downwards, are much larger and are peanut-shell shaped. Once an egg has been laid it will hatch after three days into a larva or grub. It is during the larval or grub stage that the workers select and feed any potential baby queen larvae large amounts of a high-quality protein-rich diet of a substance known as royal jelly. Royal jelly is produced by the hypopharyngeal glands of worker honey bees. The amount of royal jelly fed to larvae determines whether it becomes a worker or a queen. Worker larvae are fed royal jelly for the first three days of their life whereas larvae selected as potential queens are fed royal jelly throughout the entire larval stage.

SWARMING

It takes about sixteen days after a queen egg has been laid for an adult queen to emerge unaided from her capped cell. Two or three days before the new queen emerges from her cell the older bees in the colony get ready to swarm. During this period the worker bees who are going to swarm gorge themselves with honey and mill around the outside of the hive before flying off with the old queen to a temporary landing place, say on a tree or on a building, usually less than a few hundred metres from the hive. When the swarm has settled at their temporary home, approximately 5 per cent of the scout bees start looking for a new permanent home for the swarm to move to. It is the older bees within the swarm that tend to become scout bees looking for a suitable new nest site. In Australia, about 90 per cent of all swarms that have not been captured by a beekeeper manage to find a new permanent home protected from bad weather. The remainder die, either shortly after swarming perhaps because they have been caught by heavy rainfall in an open area or because they do not find a suitable nest to protect themselves over winter.

The first of the queens to emerge from her cell will eat some honey, gain strength and will then attempt to locate the other queen cells, chew holes through the cell walls, and sting the remaining queens to death before they can emerge and challenge her. She is aided in her search for rival queens by the peculiar piping sound unhatched queens make shortly before they emerge. The new queen is not always successful in killing the remaining unhatched queens because if she was there would not be any after-swarms, headed by other virgin queens. In fact, after-swarms are frequently reported by beekeepers.

1.12

1.12: Swarming and supersedure queen cells are located at the bottom of frames and have approximately the same shape, although supersedure queen cells are often stockier than swarming queen cells.

The newly emerged virgin queen will remain in the hive for about a week to ten days before her body has developed and strengthened sufficiently for her to go on her nuptial or first mating flight. Mating takes place at drone congregation areas (DCA) described later in this chapter.

The first swarm of the season, often called the prime swarm, may contain thousands to tens of thousands of worker bees and the old queen. Sometimes the colony from which the initial swarm originated may quickly swarm again and this is called an after-swarm. After-swarms — or secondary swarms, as they are sometimes called — are headed by a virgin queen and their departure will leave the colony even further depleted of workers. It occasionally happens that a colony may swarm a third time and this is referred to as a tertiary swarm and leaves behind very few workers in the original colony. Both secondary after-swarms and any tertiary swarms may contain insufficient numbers of workers to survive over the winter. The yearly swarming cycle of the colony will be repeated the following year as both new and old colonies increase in strength and then swarm again.

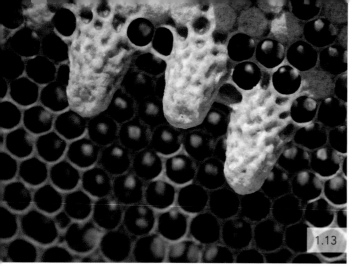
1.13

1.13: Emergency queen cells are usually found distributed around the centre of a brood frame among young brood.

1.14: Drone pupae that have not yet hatched.

1.15: Worker larvae ready to be capped. Capped worker cells are on the right.

1.14

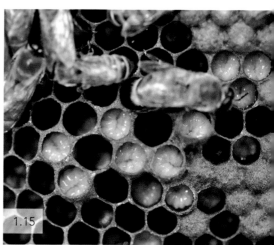
1.15

SUPERSEDURE

Supersedure occurs when an old queen is not perceived by the workers as being healthy or is not laying a sufficient number of eggs to keep the colony strong. Similar to the queen-rearing process used for swarming, with supersedure the workers encourage the queen to lay eggs in newly constructed queen cups and from one of these queen cells a new queen will emerge in sixteen days. The difference between supersedure and swarming is that with supersedure the old queen and the new queen both remain together in the colony for some period of time before the old queen dies, possibly of natural causes or due to physical eviction by workers from the hive. Outside of the hive, with no defences or ability to feed herself, she will very quickly die or be eaten. In the case of swarming the old queen permanently leaves the colony a few days before the new queen emerges from her cell.

The immediate cause of supersedure is probably diminished pheromone production by the older queen. The reason why the production of queen pheromone decreases is not clear but is more likely to occur with older queens than with younger queens.

The workers detect the lowering level of queen pheromone and start the supersedure process. The process by which workers initiate both swarming and supersedure is the same: the workers build queen cells usually at the bottom of brood frames and the old queen lays eggs inside them. The remaining development of the supersedure queen is exactly the same as for a swarm queen.

EMERGENCY

The production of queens by swarming and supersedure are events planned in advance by the workers and enable the queens to emerge in as healthy a state as possible. This process also helps to ensure that they produce strong, viable colonies as a result of swarming or supersedure.

In the case of emergency queens, however, these are formed when the queen dies suddenly and the normal swarming or supersedure process has not been started. In this situation the workers realise, after only a few hours, due to the lack of queen pheromone in the hive, that there is not a queen in the colony and they will choose some eggs or very recently hatched larvae to nurture into queens. The younger the larvae chosen to replace the missing queen, the healthier will be the new queen produced by this process. This is because older larvae will have less time to feed on royal jelly before their queen cell is capped.

Emergency queen cells may be located anywhere on the frame although they are usually not found at the bottom of the frame. They are produced by widening pre-existing worker bee larvae cells. Emergency queen cells are different to swarm or supersedure queen cells in that the workers select pre-existing horizontal worker larvae cells and then modify them so that the larger queen cell droops down from them. Thus emergency queen cells can easily be recognised from swarm or supersedure queen cells. A common reason that an emergency queen needs to be raised is because the beekeeper has accidentally or deliberately killed or removed a viable queen from the colony.

DEVELOPMENT OF WORKER AND DRONE LARVAE

In contrast to the larval life of the queen, worker and drone larvae are only initially fed high-quality royal jelly. After about three days they are fed a less nutritious diet of bee bread, which is a mixture of pollen, nectar and saliva produced by worker bees for consumption by larvae. It needs to be stressed that when the queen egg is laid there is no genetic, physical or biochemical difference between a female worker egg and a female queen egg. The only known reason for an egg to develop into a queen rather than into a worker is that the larvae selected to become queens are fed copious amounts of high-quality royal jelly during their entire larval feeding stage.

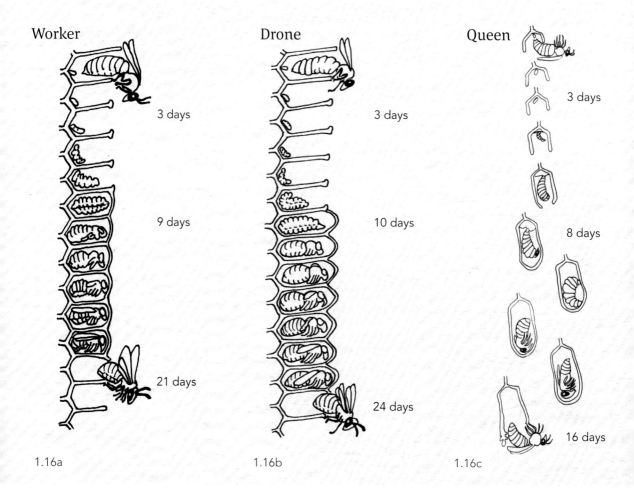

Worker

3 days

9 days

21 days

1.16a

Drone

3 days

10 days

24 days

1.16b

Queen

3 days

8 days

16 days

1.16c

1.16: Stages in the capping of brood cells.

1.16a: Development of a female worker egg, through to larval and pupal stages, and finally emerging on the 21st day as an adult worker bee. The young bee remains inside the egg for approximately three days before hatching as a larva. The larva grows until, on about the ninth day, its cell is capped ready for the larva to change into a pupa.

1.16b: Development of a male drone egg, through to larval and pupal stages, and finally emerging on the 24th day as an adult drone bee. The young bee remains inside the egg for approximately three days before hatching as a larva. The larva grows until, on about the tenth day, a day later than for a female worker larva, its cell is capped ready for the larva to change into a pupa.

1.16c: The development of a queen, through to larval and pupal stages, and finally emerging on the 16th day as an adult queen. The young bee remains inside the egg for approximately three days before hatching as a larva. The larva is fed large amounts of high-quality royal jelly and grows until, on about the eighth day, a day sooner than for a female worker larva, its cell is capped ready for the larva to change into a pupa.

MATING AND DRONE CONGREGATION AREAS (DCA)

A queen will mate in mid-air, between 10 metres and 30 metres above the ground in areas called drone congregation areas or DCA. It is believed that queens and drones select a site to be a drone congregation area based on the visibility of key landmarks such as a large tree or open spaces.

Drone congregation areas are spaced approximately every 2 kilometres apart and drones from many different colonies fly to an area for the opportunity to mate with a queen. Apart from a shorter, stubbier abdomen and absence of a stinger, the distinguishing feature of a drone is his large eyes. These have evolved to see a distant queen when she would otherwise be just a spot on the horizon. The drone's larger compound eyes also give him a wider field of vision. The competition among drones to mate is intense and there is literally a line-up of drones flying behind every queen in a fast-moving comet tail, waiting for the current drone that is mating to finish and fall off.

When a drone starts mating, an endophallus (a penis-like structure that remains inverted inside the drone's abdomen prior to mating) is inserted into the queen's sting chamber, inside of which is the queen's vagina. The drone then falls backward causing the body fluids inside his abdomen to be pushed into his endophallus, causing it to further elongate and harden. Sperm is released into the queen and the drone falls further backwards causing the tip of his endophallus to tear off from his abdomen, where it remains temporarily inserted in the queen's sting chamber. With his endophallus and many of his internal organs missing or severely damaged, the drone falls to the ground and dies. The last drone's endophallus that is still attached to the queen when she returns to the hive is often referred to by beekeepers as 'the mating sign'.

If the virgin queen does not mate with a sufficient number of drones on her first mating flight she will go on second and third mating flights to ensure a varied supply

1.17

1.17: Drone and queen mating in mid-air. The male's endophallus, equivalent to a penis, snaps off during mating and the drone falls to the ground to die.

of drone sperm from multiple colonies. A typical queen mates with between ten and twenty drones during her nuptial flights. It is believed that a virgin queen will continue to go on mating flights and mate with drones until her oviduct is full with sperm.

Queens do not mate inside their own hive, only at drone congregation areas away from the hive. The chances of a queen mating with a drone from her own colony are small because at any drone congregation area there will be hundreds of drones from other colonies. Also, studies indicate that drones and a virgin queen from the same colony probably go to different drone congregation areas to mate, further minimising the likelihood of inbreeding within the colony.

Honey bees belong to the same order of insects as ants, wasps, bees and saw flies. Most of the insects in this order are unusual, although not unique, in that although female eggs are fertilised with stored drone sperm and have both a mother and a father, male eggs are not fertilised with sperm so have no father, only a grandfather.

The queen, using her long, thin abdomen and ovipositor (a modified stinger) places the egg perpendicular to the bottom of the cell. After about a day, as the egg develops, it will soften and fall over inside the cell. The outer membrane of the brood egg is a soft membrane, unlike the hard shell of a chicken egg, and the outer membrane will slowly dissolve over a couple of days allowing the larva, which is fully formed by day three inside the egg, to emerge. During the larval stage the nursery bees will feed larvae royal jelly, although feeding will stop once the brood cell has been capped. The larvae will eat this food and will grow over the five days before their cells are capped with a porous wax cover ready for the final stage of the brood's development into pupae. It is during the pupal stage that the brood's body finally develops into a form recognisable as a young bee. At the end of the twelve- to fourteen-day-long pupal stage the young female bee chews away the wax cap of the cell from the inside and pulls herself unaided out of the cell to emerge as a young adult bee.

If the queen has laid drone eggs in the larger-cell-sized drone comb constructed by the worker bees these larvae will be attended to in the same way by nursery bees. The difference is that whereas both queens and worker bees will have finished growing after five days drone larvae will take one further day before their cells can be capped. The drone also has a longer pupal stage: a worker will emerge from her cell at around 21 days after egg lay whereas a drone will not emerge until around day 24. It will be fed by nursery bees or will help itself to stored supplies and gain strength until it begins short orientation flights in preparation for its future role of mating with virgin queens in drone congregation areas.

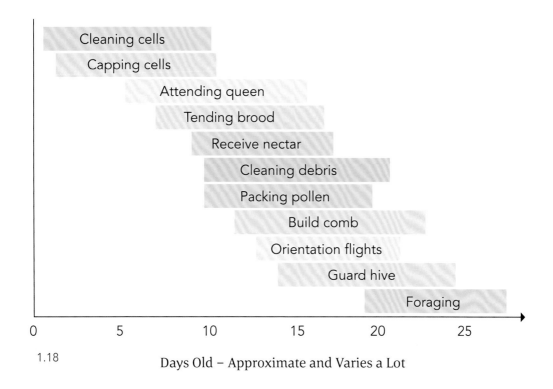

1.18

Days Old – Approximate and Varies a Lot

1.18: Stages in work roles of a worker.

LIFE AS A YOUNG ADULT BEE

Visual 1.18 shows the various stages that a worker bee transitions through from the time she first emerges from a cell to her final role as a forager collecting pollen, nectar and water. As a newly emerged adult the young bee needs to eat pollen, in the form of bee bread to obtain protein as her body still needs to grow and mature. This is stored in cells ready to be eaten. Bee bread is a mixture of pollen, honey, glandular juices and enzymes to promote the longevity of this food source. The five- to nine-day-old adult nursery bee will continue to eat bee bread to turn it into royal jelly. Royal jelly is made in the hypopharyngeal gland in her head and is fed to the developing larvae in the uncapped cells so that they too will mature and grow into adult bees.

The progress from one role to another as the worker matures is not well defined and varies both from bee to bee and on the particular needs of the colony. Not every worker performs every task. Perhaps the most dramatic change in role is from activities based solely within the hive to working outside the hive when she becomes a forager collecting food, water and propolis for the colony. Propolis, sometimes called bee glue, is a sticky resinous substance collected by foraging bees from resin-producing plants. The colony has many uses for propolis including blocking holes, cementing and strengthening cell bases, even embalming small animals that have died in the nest such as mice that are too large to be removed by the bees.

This dramatic shift in worker behaviour correlates with a change in protein levels within the body of worker bees. Of particular interest is a protein called vitellogenin, which makes up 30 to 50 per cent of the protein present in a bee's haemolymph (insect blood). Vitellogenin is used in the production of the royal jelly produced by workers for feeding to larvae. In addition to it being an ingredient in the food produced by nursery bees, vitellogenin has beneficial effects on the health of the bee, assisting in maintaining immune system functioning as well as in suppressing the production of juvenile hormone, a hormone that promotes the ageing process in insects. With the switch from nursing behaviour to foraging, workers greatly reduce their production of vitellogenin, leading to their ageing and eventual death. Older forager bees can return to nursing duties if required by the colony but do not fully return to the levels of vitellogenin production seen earlier in their lives for the reason that their pharyngeal glands would have reduced in size when they initially became foragers.

An adult worker bee lives for about 30 to 50 days during the spring and summer and may live considerably longer if born during the months preceding a cold winter. During the winter months in colder climates brood production may cease altogether with the adult bees huddling together on the comb in order to keep warm.

The drone has no elongated tongue or proboscis to suck nectar from flowers and with his almost hairless legs he has no useful apparatus to gather pollen. He is reliant on the worker bees in the hive to provide him with food and as he has no sting is unable to defend the colony during times of attack.

1.19: Developmental differences between workers, queens and drones relating to DNA and feeding. Both the worker and the queen result from an egg fertilised with the sperm from a drone. The only development difference between a female worker and a queen is that the queen is fed far more high-quality royal jelly during its larval stage. The male drone only has its queen mother's DNA since its egg is not fertilised by male drone sperm. The food fed to the male drone larva is the same as fed to a female worker larva.

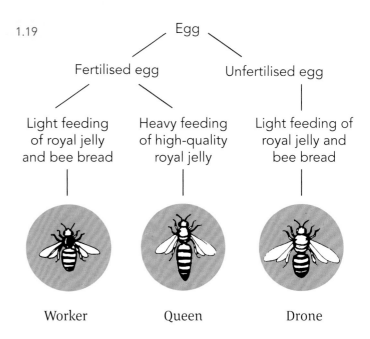

1.19

Egg

Fertilised egg — Unfertilised egg

Light feeding of royal jelly and bee bread — Heavy feeding of high-quality royal jelly — Light feeding of royal jelly and bee bread

Worker — Queen — Drone

Although there is ongoing research and some argument among beekeepers, the current wisdom is that the only function of a drone is to mate with a queen and that the male bee in the hive has no other role in support of the colony. It is not surprising then that when winter approaches and stores of honey and pollen must be preserved the workers evict the hapless drones and simply raise more the following spring when they are needed for mating.

COLLECTING NECTAR AND POLLEN

ORIENTATION FLIGHTS

Before a worker bee can start foraging it must first learn to recognise the area around the hive so that it can use landmarks to safely return home. Since up to 2000 eggs a day are laid, a lot of potential foragers need to take their initial orientation flights every few days. The first orientation flights are short practice flights and enable a bee to memorise the location of its hive and any nearby landmarks to use for future reference. These flights can sometimes be recognised when there are hundreds of bees flying seemingly aimlessly outside a hive. Some people mistake these for robber bees but robber attacks are much more aggressive with a fast darting motion toward the hive entrance. Orientation flights are also often mistaken for swarming flights but careful observation will show that the bees are circling and moving near the hive whereas a swarm will, after a few minutes, leave the hive and move off. Drones must also orientate themselves to the position of the hive and begin taking short orientation flights when they are approximately eight days old.

FORAGING

Foraging bees use their proboscis or mouth parts to suck nectar from flowers or water into a sac or honey stomach inside their body, known as a crop. When the crop is full of nectar or water the foraging workers return to the hive where the food is passed to specialised in-hive workers whose sole job is to take the nectar or water from the returning bee and to store it in empty cells. The process of mutual feeding by social insects or taking nectar or water from a returning forager bee is called trophallaxis. With trophallaxis, the returning forager brings up nectar from her crop into her mouth where it is taken by the in-hive worker. The regurgitated nectar is then stored in the crop of the in-hive worker where it is mixed with additional enzymes. The in-hive worker with the new supply of nectar will then take it to a honey cell for storage.

1.20: Bees feeding each other. One bee contains a store of nectar or honey in its sac and is regurgitating it and feeding it to the other two bees. The scientific name for this process is trophallaxis.

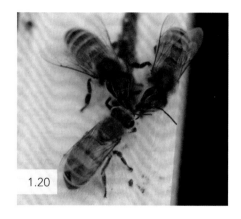

1.20

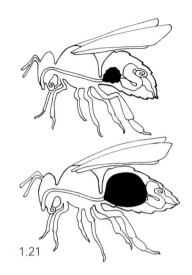

1.21

1.21: The honey crop is the sac inside a worker's abdomen that is used to store nectar for transport from the flower back to the hive. When the sac is full of nectar its volume increases and takes up much more room inside the forager's body. The above two drawings show the honey crop both full and empty.

Nectar has a high water content, often greater than 60 per cent and this needs to be reduced to about 17 per cent before the nectar can be called honey. When the honey has the correct water content the cell that it is placed in for storage is capped until it is needed for food by the colony. Evaporating water alone will not cause nectar to turn into honey and it is when the nectar is in the foraging bee's crop that it is mixed with enzymes to convert the sucrose in the nectar to fructose and glucose as well as causing other chemical changes to take place.

If you look inside a hive when there are many types of plants and trees flowering you will see a kaleidoscope of pollen colours inside the cells around the brood. This is because when a forager collects pollen it practises flower fidelity and only collects one type of pollen during a flight, flying from flower to flower of the same type of plant or tree. The foraging worker will become covered in pollen at the flower and uses her front and middle legs to clean her body of pollen and to store it on her specially adapted hairs on her back legs. This grooming is performed in the air shortly before she flies to the next flower or returns to the hive. While she is in the field the foraging bee mixes the collected pollen with honey stored in her crop for this purpose. This mixture forms a sticky paste that will attach itself more easily to her hind legs. It is easy to observe returning bees at the hive entrance with their two compressed pollen pellets attached to the outside of their rear legs and often you will also see excess loose pollen attached to the hair on their bodies.

COMMUNICATIONS AND NAVIGATION

THE DANCE LANGUAGE OF HONEY BEES

In order for a complex colony of bees to survive they need to communicate with each other and to coordinate many of the colony's activities, particularly those relating to the collection of food. This enables a colony of bees to achieve much greater results than if each bee were to act on its own without regard to the activities of other bees. This coordination and communication between bees is at its greatest and most complex when foraging bees return and need to tell other bees within the colony of new food resources and the food source's direction and distance from the hive. These activities can be broken down into two separate but interrelated functions.

1.22: Foragers store pollen on their rear legs to carry it back to the hive. Using the front two pairs of legs they brush pollen from their body, mix it with saliva and nectar or honey from their crop to make it sticky, and then attach it to their back legs.

1.23: Pollen stored in a frame. Before storing pollen the bees mix it with honey and salivary juices containing enzymes so that it will store better and also become more nutritious.

1.24: Different sized capped cells for workers and drones. Workers have smaller flat-capped cells while capped drone cells are larger and protrude above the surface of the comb.

1. A returning forager needs to inform other bees within the colony of a new food source and its direction and distance. This is done using the waggle dance.

2. Foraging bees leaving the colony must be able to navigate independently to the food source and then find their way back unaided to the hive.

THE WAGGLE DANCE

In order for a returning worker bee to communicate to other bees within the colony the location of new nectar, pollen, water or propolis resin sources, the returning bee performs a waggle dance. The waggle dance indicates the direction and distance of the new food source from the hive.

Movement in the waggle dance consists of a bee running straight ahead for a short distance, up to about 1.5 centimetres, making a buzzing noise and energetically shaking vigorously, or waggling, her abdomen from side to side. At the end of the first straight run the bee turns in one direction to return to the start of the straight run. She then performs another straight run shaking her abdomen vigorously, and then at the end of the run turns in the other direction to return to the start of the straight run. This dance pattern draws out a figure of eight and, as a result, the waggle dance is often called a figure-of-eight dance. The waggle dancer periodically stops dancing to allow

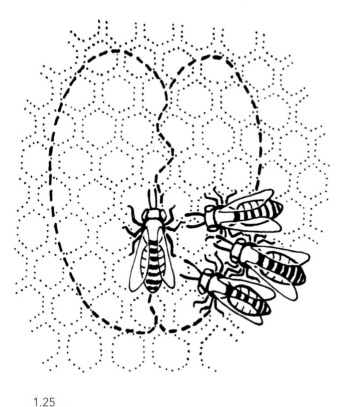

1.25

1.25: A forager bee doing the waggle dance inside the hive. The bee performing the dance is telling the watching foragers the direction and distance of the new food source. Periodically the watching foragers will stop the dancing bee and stroke her with their antennae. This is believed to provide the watching foragers with information about the scent of the new floral source that they can use to further refine their search for the food source.

1.26: The orientation of the waggle dance to the vertical inside the hive tells the bees that are watching the direction relative to the sun of the new food source. If the dance is vertical, the food source is in the direction of the sun. If the dance is at an angle of, say, 40 degrees to the right of vertical, the food source is at an angle of 40 degrees to the right of the direction of the sun from the outside of the hive.

the watching workers to touch her with their antennae, probably to detect the scent of the food source.

The most significant aspect of the waggle dance is that it conveys to the watching forager bees both the direction and the distance to the food source from the hive. Direction to the food source is given by the angle off vertical that the returning worker aligns the straight waggle part of the semi-circle to. If the straight waggle part of the dance is performed at an angle of 40 degrees to the left of vertical, the direction of the food source is 40 degrees to the left of the direction of the sun from the hive.

Distance to the food source is given by several means. There is the tempo of the abdominal waggle that decreases with greater distance to the food source from the hive. There is also the length in millimetres as well as length of time taken to perform the straight waggle portion of the dance which lengthens with increasing distance to the food source.

The accuracy of the dance is constrained by the speed and distance over which a dancing bee is able to pirouette back to her starting position during the return phase of the dance.

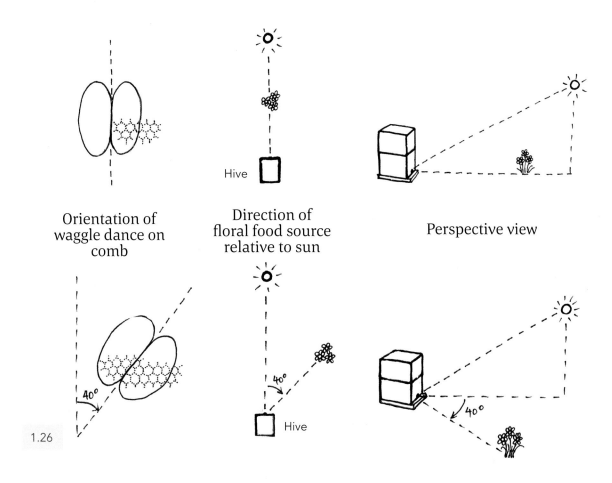

Orientation of waggle dance on comb

Direction of floral food source relative to sun

Perspective view

1.26

It used to be thought that returning forager bees used a simpler type of dance, the round dance, to tell other bees about a food source less than 100 metres from the hive. More recent studies have shown that all waggle dances follow a figure-of-eight pattern. Due to space constraints within the hive this pattern is sometimes difficult for both the bee to perform and for the beekeeper or scientist to identify.

NAVIGATION AND ORIENTATION DURING FLIGHT

Although the dance language provides information to forager bees about the direction and distance to food sources, it does not provide a mechanism for the forager bees to navigate during flight. Like all animals, the honey bee uses a variety of complex information to provide both navigation and orientation information during flight.

Honey bees primarily use the sun and polarised light from the sun to provide direction information. The use of polarised light enables the bee to navigate by inferring the location of the sun when it is not visible (for example, by trees or partial cloud cover). Under full cloud they are unable to use polarised light and must navigate by landmarks. The most remarkable fact about the bee using the sun as her primary source of direction during flight is that she is able to compensate for the movement of the sun as time passes. Thus, if the sun has moved 10 degrees between the time a bee leaves a colony in search of food and the time she starts her return flight, she will make her own correction for the sun's new position in order to use it as a guide to return home. Geographic cues or landmarks are also used by bees as a secondary tool for navigation as well as the direction of the Earth's magnetic field.

HONEY BEES AS SUPERORGANISMS

Beekeepers often think of a hive as a single living entity and, when looking inside a hive, will see thousands of individual bees going about their daily specialised tasks in support of the colony. Each member of a honey bee colony works cooperatively with the other bees to ensure that the colony survives, with each individual bee working only for the good of the whole colony.

Because of this, honey bee colonies are considered to be superorganisms as the entire colony, rather than the bees individually, is viewed as the biological unit. Each individual within the colony is unable to survive in isolation from the colony as a whole. The queen and drones require the workers to keep them fed and rear the young, while the workers themselves are unable to reproduce and require the queen to produce the next generation.

The roles that individual bees perform that are essential for survival include nursery work, capping cells, storing food, guarding the colony entrance and foraging. Just as a colony would not survive without individual bees attending to their specialised tasks, individual bees would not survive for long outside of a colony that supports them. In this way each individual bee within the colony forms an unbreakable bond with the other members of the colony.

SUPERORGANISM DECISION MAKING

Because superorganisms are made up of many individual members, the honey bee colony relies on self-organising behaviours rather than centralised control for the colony to function. These behaviours do the following:

- Use simple, repeating feedback loops. For example, when a worker stings a beekeeper she also releases an alarm pheromone that recruits other workers to the attack.
- Are used to gather information from multiple individuals. For example, a swarm scout

1.27 A colony of honey bees is called a superorganism. Although made up of tens of thousands of individual bees, the bees work together for the survival of the colony.

1.27

returning to the swarm cluster with information on a possible new nest location needs to have the nest site validated by many other scouts before the swarm as a whole will move to the new location.

- Enable communication within the colony and ensure that information is not deliberately given to rival colonies; for example, the waggle dance is performed within the hive and information on new floral sources is only passed to members of the colony.
- Use positive feedback cycles to initiate and coordinate actions by other members of the colony. For example, a returning forager uses the waggle dance to inform other foragers inside the hive of the location of a new food source. These newly recruited foragers will return from the food source and, if the new floral source is still providing nectar and pollen, will also perform the waggle dance to recruit many more foragers to visit the flowers.
- Positive feedback enables the colony to make decisions collectively by involving many worker bees in the decision so that a reliable choice is made that benefits the colony as a whole. Examples include organised nest hunting by swarm scouts and the organisation of work activities within the hive.
- Control foraging activities. For example, if in-hive bees did not immediately take nectar from a returning forager, foraging activities would be reduced. This would happen if the colony did not have sufficient empty cells in which to store incoming nectar.

In the honey bee colony it is not the instructions of an individual bee within the colony such as the queen, but rather the sheer number of bees, each with their own part to play in support of the colony, that enables the colony to survive, reproduce and the species to flourish in many parts of a hostile world.

SUMMARY

- Each colony of honey bees contains a single queen, between 20,000 and 60,000 female workers and up to 1000 male drones. All of the work and decision making within the colony is performed by the female workers. Each colony will swarm at least once each year, leaving behind about half of the old colony with a new queen. The other half of the colony will depart with the old queen to seek a new home and to establish a new colony.

- The three most common races of bee kept by beekeepers in Australia are the Italian, Caucasian and Carniolan races, although the Italian bee is the race most frequently kept.

- Worker bees live for about six weeks once they emerge from their capped cells as adults. The first three weeks as an adult are spent inside the hive performing duties such as looking after brood or cleaning. The last three weeks of their life are spent as forager bees searching for and collecting nectar, pollen, propolis and water. In comparison to worker bees, drone larvae take extra time to grow and to pupate into an adult. Both worker and drone bees take orientation flights prior to becoming foragers or going on mating flights.

- Forager bees use a dance language to communicate with bees inside the hive the direction and distance to new food sources. The foraging bees use polarised light and landmarks to navigate to both the hive and floral sources.

- A honey bee colony is regarded as an example of a superorganism as each bee in the colony works co-operatively for the good of the colony as a whole. The entity of the colony is paramount to the individuals in it.

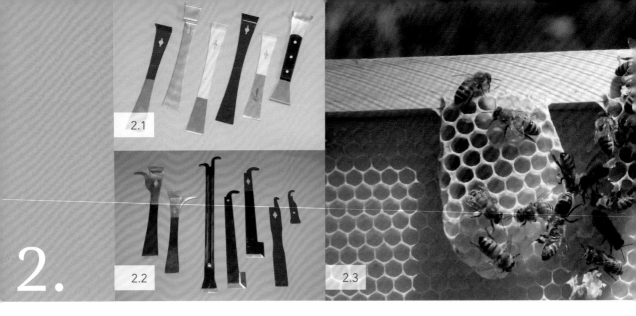

2.1

2.2

2.3

Basic equipment and protective clothing

New beekeepers usually have two initial questions:
- What equipment and protective clothing do I need?
- Where do I get my first hive of bees?

This chapter will give you an idea of the minimum amount of equipment needed to get the hobby beekeeper started. Chapter 5: Your First Bees will cover the second question.

The new beekeeper needs protective clothing, a hive tool, a smoker and a bee brush to be equipped and ready for their new hobby. The selection of this equipment is important as there are various designs available and, in the absence of good advice from a supplier or an experienced beekeeper, deciding which of a number of choices is right for your circumstances can be difficult.

EQUIPMENT

HIVE TOOL

There are two general designs of hive tool — the Australian and the American patterns. Both have their advantages although hobby beekeepers overwhelmingly prefer the Australian design, which is often called a J-tool for no better reason than it looks like a large metal J. The J hook at the end of the tool allows for easy removal of frames that are stuck in the hive either by honeycomb or by propolis. The wide crowbar-type hook at the end of an American hive tool is harder to use for this purpose. This is particularly

true when the very first frame is removed from the hive, which when stuck fast can prove difficult to raise as the gap between the frames is too narrow for the beekeeper to get a good grip on the top of the frame. At the opposite end to the hook both types of hive tool have a sharp, chisel-like edge that is used to prise open lids, wedge apart frames and to loosen boxes adhered together by the bees. This edge is also useful to clean burr comb from frames, lids and boxes. Burr comb is the small pieces of wax honeycomb that worker bees build away from the central part of a frame, such as on the walls of the hive or underneath the lid.

When choosing a hive tool check that the chisel edge is reasonably sharp and can easily be levered underneath a lid to prise it open. Some tools available on the market are too thick to work efficiently for this purpose. Many professional beekeepers prefer the American design because if they need to prise open dozens of hives every day, the flatter hook at the other end is more easily hit repeatedly with the palm of the hand without sustaining injury. It is common for a beekeeper to own two hive tools as these are inexpensive to purchase. An added advantage of owning two is that when one is mislaid, as it almost certainly will be, the beekeeper has another tool to use during hive inspections.

A supplier will usually recommend the standard Australian J-tool to a new beekeeper; however, within the two general patterns of Australian and American hive tools there are many different styles and designs — some tools are longer while some are shorter. As the beekeeper gains more knowledge and skill it is worthwhile considering an additional design of tool.

If you suspect that a colony that you have just inspected is diseased, to reduce the probability of spreading the infection to other colonies, your hive tool can be cleaned

2.4

2.5

2.6

2.4: If you believe that a colony that you have just inspected is diseased, a quick and effective way to sterilise your hive tool is to leave it in a smouldering smoker.

2.5: Some smokers use the older design with a grille insert in the smoker.

2.6: Other smokers use the grate insert. Once the smoker has been lit it is much less likely to go out with this design of grate.

by placing it inside a burning smoker. Alternatively the hive tool can be cleaned by leaving it in a jar of strong bleach or disinfectant.

SMOKERS

Smokers are an essential part of the beekeeper's equipment. Filled with smouldering fuel to produce copious amounts of white smoke, the smoke is puffed over angry bees to calm them down. When bees smell smoke their instinctive fear of fire causes them to gorge on stored honey supplies. They are generally then more docile and have difficulty stinging, as they are unable to bend their overfull abdomens. On hot days when there is a shortage of food or because the bees were born with an aggressive streak, the beekeeper inspecting a hive is sometimes confronted by a mass of defensive bees. A few puffs of smoke will calm even the most agitated colony, allowing the beekeeper to inspect the hive in more comfortable conditions.

In times past the majority of smokers used by beekeepers were made of galvanised steel. These days, better quality smokers are made of stainless steel and various models are available to suit every beekeeper's pocket and preference. Smokers vary in size from small to large with each having its own advantages and also a few disadvantages. Many beginners turn to an experienced beekeeper for advice on which type and size of smoker to buy. Some will be told to buy the largest available as it will hold plenty of fuel and will not need topping up during hive inspections. There are also a number of beekeepers who swear by a small smoker that is a little larger than a soup can. Generally most suppliers will recommend a medium-sized smoker with dimensions approximately 25 centimetres to 27 centimetres high and 10 centimetres in diameter. This size is not too heavy to hold and has a reasonable fuel capacity to work a small number of hives. A commercial beekeeper with large numbers of hives to inspect may prefer a larger smoker with a greater fuel-holding capacity.

For the hobby beekeeper with only a few hives a medium-sized smoker is the most convenient one to use. When purchasing a smoker check that on the outside there is a protective cage or cover to prevent the hot cylinder touching you and causing burns. There are two types of lids on the majority of smokers sold. One is usually made of a single rolled piece of stainless steel which ends in a pointed spout — this is often referred to as an American style of smoker. The other commonly found shape is a dome with a welded spout. This style is generally referred to as a European, German or domed smoker. Keep in mind that the shape of the lid is purely cosmetic and that the other features of the smoker should also be considered.

One of those considerations is the design of the internal grille. This is used to distribute air and to prevent the fuel burning a hole through the base of the smoker. It's a simple, flat piece of perforated metal raised off the bottom of the unit. This style of grille is commonly found in the more traditional style of smoker, but the fuel tends to stop burning if the smoker bellows are not 'puffed' or depressed every minute or so. Recently gaining more popularity is the type of cylindrical insert usually seen in the European style of smoker. This essentially is a cylinder with a large number of perforated holes in it. A bolt centrally placed in the bottom holds the cylinder up and off the bottom of the smoker. This type of grille affords a lot more aeration to the fuel being burnt and as a result the beekeeper does not need to resort to pumping the bellows quite as often. This can be an advantage for the beginner whose hive inspections take longer thereby increasing the risk of either the smoker running out of fuel or going out because the beekeeper is concentrating on what is happening inside the hive and not on the smoker outside. There are advantages in a smoker that will continue smoking with minimal attendance as there is nothing worse than realising your smoker has gone out at a critical moment, for example, when you are covered with bees, which will require you

2.9

2.10

2.13

2.7: Lighting a smoker. Put crumpled newspaper into the smoker and light it using a barbecue lighter.

2.8: When the paper is well alight keep puffing the bellows and slowly add dry pine needles until white smoke is pouring out.

2.9: Puff the smoker until there is plenty of white (cool) smoke coming out.

2.10: Putting dry gum leaves into a smoker to use as fuel instead of pine needles.

2.11: Putting a roll of corrugated cardboard inside a smoker to use as fuel. Rolled inside the cardboard are some pine needles.

2.12: Using a torch to light a smoker. This is a quick and easy way to light a smoker that contains any kind of fuel.

2.13: Inspecting a hive. Step 1: When applying smoke to a hive, first puff some smoke into the entrance.

2.14: Step 2: Next slightly raise the lid and puff smoke under the lid.

2.15: Step 3: Wait about 30 seconds, remove the lid and hive mat then puff smoke over the top of the frames.

2.16

2.16: A smoker needs to be carried safely inside a vehicle to avoid the hot metal causing a fire. I place the hot smoker in a metal bucket when I store it in my ute.

to move away from the hive. There is also the time involved in relighting the smoker to consider.

Generally the European smoker with the round cylinder type of grille is more expensive so this may be another consideration when deciding on which style to purchase. Keep in mind that a quality smoker will last many years with the only major ongoing repair being replacement of the smoker bellows from time to time. For this reason it is always worth asking if there are replacement bellows for the unit readily available before you buy the smoker.

There are many ways to light a smoker and every beekeeper has a personal preference for how to do this and for the fuel that is used. A barbecue lighter is an excellent way to provide the flame to light the smoker fuel as it has a long nose suitable for pushing deep into the fuel after it is placed in the smoker. First, some newspaper is either crumpled or shredded and loosely balled into the bottom of the smoker. Flame is applied to light the paper and then the bellows are gently puffed or depressed until the paper is alight. At this time the fuel of choice is added until it is well alight and smoking profusely. My personal preference is for dry pine needles, although paperbark, dry gum leaves, hessian sacking, sugar cane mulch or rolled up cardboard can also be used as fuel. Once the smoker is well alight add more fuel and press it down into the smoker, puffing the bellows all the time. Once this is complete and white smoke is pouring profusely from the smoker it is ready to use and the lid can be closed. A note of warning: please do not approach your bees while the smoker is igniting as bees require 'cool white smoke' rather than hot smoke.

An alternative to the barbecue lighter is to use a propane torch sold in hardware shops for brazing metal pieces together. These torches are easy to light with the click of a switch on the handle and their intense heat will very quickly get smoke pouring out of the smoker.

It is a good idea to keep a fireproof tin, bucket or empty ammunition box to place the smoker in after use if you are to travel some distance to and from your hives in a vehicle. Many a sad story has been told of bee suits and equipment catching alight in the back of vehicles due to carelessly carried burning smokers.

A smoker is dangerous if left unattended and should be completely extinguished

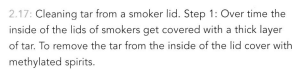

2.17: Cleaning tar from a smoker lid. Step 1: Over time the inside of the lids of smokers get covered with a thick layer of tar. To remove the tar from the inside of the lid cover with methylated spirits.

2.18: Step 2: Set fire to the methylated spirits and let it burn out, being careful that the bellows don't also catch alight.

2.19: Step 3: Once the methylated spirits has burnt out the tar will have become brittle and can easily be removed using a hive tool.

2.20: A well-loved smoker.

after use. Careful attention should be paid to where any hot embers or smoking fuel remains are discarded and water should be used to ensure the remains cannot reignite. It is a good idea to leave the embers to extinguish in the smoker and to check that they are completely cold prior to tipping them onto the ground. Smokers are not permitted on total fire ban days. Many beekeeping suppliers sell a liquid smoke that can be mixed with water and applied using a spray bottle for use on such days.

Over time smoker lids get clogged up with heavy deposits of tar. The tar can reduce the size of the smoker nozzle making it less effective to use as well as making the lid difficult to close. To remove the tar from the lid, pour some methylated spirits over the inside of the lid and set it alight. Make sure that none of the methylated spirits is on the bellows or any wood or plastic parts of the unit. Once the methylated spirits has burnt out, the tar will have become brittle and can easily be broken off using your hive tool.

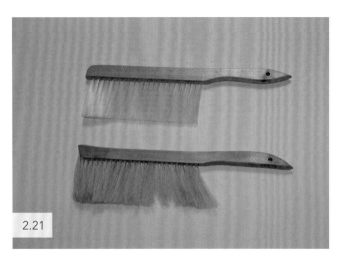

2.21

2.21: Bee brushes are an essential part of a beekeeper's toolbox. Most brushes are either made with nylon or horsehair bristles. Nylon is more durable although horsehair is gentler on the bees.

BEE BRUSHES

Although it may not be apparent to a new beekeeper, a bee brush is an essential and inexpensive item of equipment. A brush comes in handy when the beekeeper wants to remove bees from frames or off the rim of the hive box to replace the lid. Nylon or horsehair brushes are usually used. Nylon is easier to clean. Horsehair is softer but it is also prone to splaying out and may trap bees. After using the brush, rinse it in cold water to remove any honey or other residue and the brush should last for years. The brush needs to be thin to stop bees and debris getting caught in the hair, and the width or thickness of the brush should be about 1 centimetre or less. Just like a hive tool a bee brush can transmit disease from one hive to another.

PROTECTIVE CLOTHING

Every beekeeper needs to protect his/her face as stings on the eyeballs, although extremely rare, may cause temporary blindness. Stings on other parts of the face, particularly the forehead, can cause significant swelling. With this in mind there are three types of protective clothing that can be used.

- Veils that protect only the face.

- A jacket with a built-in veil that offers good protection from the waist up.
- Full-length overalls that offer complete protection for the whole body.

VEILS

Veils vary in design both in the way they protect the wearer's head and also in the way they are secured to the wearer's body. Some veils come without a sewn-in hat and these are worn over either a beekeeper's helmet or your hat of choice. The top of the veil is elasticised so that it fits over the crown of the hat. Over time the elastic can become loose and offer an entry point for bees. A veil with a hat sewn into it avoids this problem. Many hats have metal grilles around the face while others are made of nylon or polyester netting. The bottom of the veil is made of fabric or netting, which will either have a long string attached for tying around the wearer's torso or, alternatively, an arrangement that fits over the shoulders and ties or is elasticised under the arms.

While relatively inexpensive, veils have the significant disadvantage that it is often difficult to prevent bees gaining entry under the bottom of the veil and up into the face area. This is not generally a problem though, as the trapped bees are usually more concerned with getting out of the front of the veil to freedom. The psychological consequences, however, of having one or more bees inside a veil can be quite devastating. Again, the beekeeper will be required to weigh up the costs involved in starting with the cheaper veil and whether these compensate for the extra vulnerability to stings. My opinion is that veils do not provide the same level of

2.22a and 2.22b: Veils are inexpensive and easy to put on. The disadvantage is that they offer less protection than a jacket or overalls since bees are able to get inside a veil more easily than with other types of protection. This is overcome with the design of veil in 2.22b with the pullover style top.

protection as a jacket or overalls, although they are cheaper and many beekeepers use them, particularly professional beekeepers. It is critical for the beekeeper just starting out that a level of confidence is gained while working with bees and this will be boosted by the extra protection offered by a jacket or overalls.

The composition of the netting on veils used in all protective clothing is varied, so choose thicker nylon netting or metal gauze netting so that the veil will last longer. The mesh in the veil needs to be black since it is impossible for a beekeeper to see through a white mesh veil.

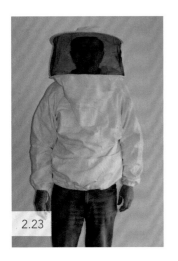

JACKETS

The most popular option is the jacket with a built-in veil. The built-in veil provides good visibility and is usually open enough to provide plenty of air to the beekeeper. Some jackets come with a zipper up the front, like a normal jacket, or may be designed without a front zipper and pull over the wearer's head. If the jacket has a zipper up the front, check that the join where this reaches the veil does not have a gap that will allow bees to enter.

2.23: This popular design of overalls uses a round hood supported by a round wire band and allows for plenty of ventilation and light. Note that there is no zipper on the pullover. The round metal band ensures that the protective netting cannot come in contact with the face.

2.24: This style of jacket uses a fencing-style of hood.

Good-quality jackets will have a Velcro patch where the jacket and attached veil zippers meet to ensure that bees cannot enter the jacket under the chin. Some manufacturers have gone further and designed the veil so that the zippers actually cross over each other in the front and are further sealed with a Velcro tab over the top. When selecting a jacket (or overalls), take particular care to check that the veil can be removed for laundering. To prolong the life of your veil it is generally unwise to launder it in a washing machine, as the abrasive action will destroy the netting inserts. Usually the veil will not need laundering as frequently as the jacket, and when required, this is better done by hand. Another important consideration is how easy it is to zip the veil closed when you are suited up. If there is only one zipper attaching the veil to the jacket it may be difficult to reach around the neck to grasp the zipper head when closing the hood. Always try on any jackets or overalls you are thinking of buying to make sure you can get in and out of the clothing easily.

When wearing the jacket make sure that the elasticised bottom around the waist forms a seal against bees. The jacket should remain above the crotch because if it is pulled down further any gap formed between your legs will quickly be taken advantage of by bees in the vicinity. Try to buy garments with good quality, well-known branded zips. Jackets are about twice as expensive as veils, but the extra cost is well worth the added protection that they provide over a simple veil. This is particularly so for the beginner.

A good jacket or overall should have elastic loops sewn on both wrist cuffs. These hook over the thumbs to stop the sleeve moving up the arm, which can result in stings on the forearm or wrist if you are gloveless when working with your bees. The loops are also handy to prevent sleeves riding up when gloves are pulled on.

OVERALLS

Full-length overalls with a built-in veil are the most expensive option but do provide the best protection for the beekeeper. A big advantage of overalls is that they give you full protection and confidence when dealing with bees. The overalls should come with elasticised wrists and ankles to stop bees from crawling inside. Like the

2.25

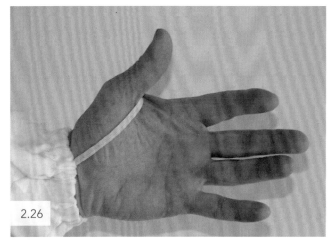

2.26

2.25: The part where the zipper joins the hood and the jacket is a critical part of the design. To minimise the chances of bees entering the clothing make sure that the two ends of the zipper overlap and that there is also a Velcro cover over the zipper.

2.26: Elastic bands are often attached to the wrist to place over the thumb of the beekeeper. The elastic bands stop the sleeves from riding up the arm.

jacket, there should be elastic loops at the wrist to put over your thumbs to stop the sleeve moving up your arm. Again check the zippers and make sure that the fit is comfortable with plenty of room to move, particularly around the crotch area. The crotch to high shoulder point measurement is the key to a good fit in overalls, so don't rely solely on chest size, particularly if you buy online. There should be no restriction when the arms are raised over the head or when bending over with the arms placed near to the ground.

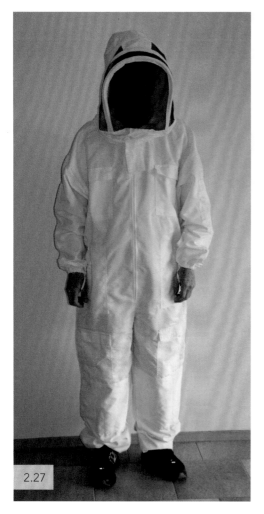

2.27

2.28

2.27: Full-length overalls provide complete protection to the beekeeper and usually come with elasticised ankles as well as wrists.

2.28: Protective clothing made out of aerated fabric is much cooler to wear on hot days. The fabric is made of three layers of material: an inner and outer shell made of nylon and a mesh material in the centre that provides thickness to the clothing as well as allowing for air flow.

In the summer beekeeping is hot work and it is best to get a very loose fitting jacket or pair of overalls that allow air to flow around the body and provide flexibility of movement. The material that the jacket or overalls are made of is important so try to buy a polyester/cotton blend that will not shrink when washed. As a precaution, ask the supplier if the cloth in the overall or jacket is pre-shrunk. Generally the polyester content will ensure the longer life of the garment particularly at high-wear points such as the elbows or knees. It is possible to get pure cotton clothing, but again be aware of the possibility of shrinkage.

A jacket or a pair of overalls purchased in the mid-price range should give the beekeeper many years of service. Keep in mind that the veil will usually need replacing during the life of the garment and you should check with the supplier if replacement veils are available. If not, you may need to occasionally repair a veil, as these are usually the first part of the garment to deteriorate. Veil repair is discussed later in this chapter.

AERATED FABRIC

The standard garment made of either cotton or a polyester/cotton mix with long sleeves and long legs in the case of the overalls does mean that beekeepers can become very hot even in mild weather. A recent innovation is the newer aerated garments that are now available. Even though they are much more expensive, protective clothing made of aerated fabrics keeps the beekeeper much cooler on hot days or during long hours of work. These garments are made of strong, thin outer and inner layers of fabric that are tear resistant and provide physical protection to a thicker rubberised inner layer. The rubberised inner layer is a loose weave net to allow air to circulate freely through the fabric. The inner and outer nylon layers on the surface are also net-like to allow air to circulate. The garments are see-though and due to their layers offer better protection against stings as well as being much cooler to wear than clothing made of conventional materials.

GLOVES

Most beekeeping gloves are made of cow or calf hide and have a long sleeve or gauntlet with an elasticised top to prevent bees crawling down into the glove. Gloves made of much softer goatskin may also be used. Although all of these gloves provide good protection against stings they have the disadvantage that the material is thick and finger dexterity or feeling is lost while wearing them. Some gloves come with an aerated wristband to keep the wrists cooler, but as the wristbands are frequently made of thinner material they can allow for the occasional sting.

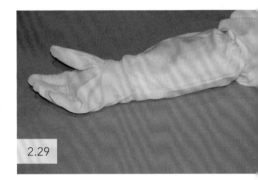

2.29

All beekeepers work to a budget and many decide to use the thicker chemical-resistant work gloves they already have at home or those that can be bought cheaply from a hardware store. These gloves provide very good protection to the hands, but the lack of a gauntlet or long sleeve means that care needs to be taken to ensure that the wrist does not become uncovered during hive inspections. Also, since chemical gloves are impenetrable to sweat, hands become very smelly after even a short use — a smell that is difficult to wash off. If plastic, rubber or nylon gloves are bought to handle your bees, buy a larger size and purchase some cotton gloves to go inside them since these will absorb sweat, are washable and will also ensure protection against stings to your hands.

2.30

2.29: The conventional design of beekeeper's gloves includes the long gauntlet and a wrist vent that keeps the hands and wrist cool. The gauntlet of the glove is worn over the sleeve of the jacket or overall.

2.30: If you use plastic gloves, wear a thin pair of cotton liners inside the gloves to stop your hands becoming sweaty and smelly.

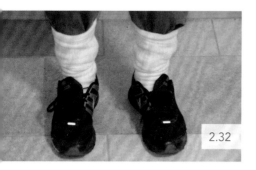

2.31: Nitrile gloves are inexpensive and provide a much better sense of touch than leather or plastic gloves. Although good for handling bees, working with queens and removing frames from hives, they are not suitable for lifting heavy hives as they may tear.

2.32: If you do not wear a full pair of overalls a pair of thick socks will provide protection to the ankles. Bees congregate around the socks and, if given the opportunity, will climb up the leg before stinging you. Make sure that your jeans are tucked into a very thick pair of socks.

An alternative to each of the above is to use Nitrile gloves. These are the gloves used by mechanics to keep their hands clean when working on cars. Nitrile gloves have the advantage that they are strong, fit snugly over the hands, and have excellent feel and dexterity. Bees have trouble stinging through Nitrile gloves although they are not as sting-proof as cow or calf-hide gloves. Nitrile gloves should not be confused with the weaker latex dishwashing gloves found in many homes. Nitrile gloves can be found at the supermarket, hardware or car accessory store, although the cheapest place to buy them is online.

My preference for beekeeping clothing is a jacket, loose jeans, gloves and very thick socks to tuck my jeans into. Inevitably the bees will zero in on the socks and try to sting through them. Make sure that you always use thick socks and that your trousers are tucked in, even on brief inspections, otherwise the bees will seize the opportunity you present to them and sting your ankles.

An alternative way to protect your ankles is to wear gaiters available from any outdoor or ex-army shop. These are easy to put on and provide almost complete protection to your ankles. Alternatively, you can wear shin-length boots with trouser legs tucked in or gumboots with trouser legs over the boots and secured with an elastic band. Overalls will prevent most of the ankle vulnerability that can be experienced by the jacket-wearing beekeeper.

CLOTHING FOR CHILDREN

Children are fascinated by bees and will often take an interest in the beekeeper's hobby. Overalls are recommended for children as they provide the most protection. Buy a little larger than is currently needed, as the child will quickly outgrow a suit that is a good fit when purchased. Always ensure that there is a good seal around the wrists and ankles. Long socks worn over overall legs with the addition of gumboots provide the best protection. Take particular care that the mesh in the veil cannot touch the child's face as a painful sting could turn the youngster off beekeeping for life. This can usually be achieved by encouraging the child to wear a baseball cap under the veil, particularly with the fencing style type of veil. Turn up the collar of the overall inside the veil as added protection. To make the overalls more attractive to the

2.33: Children love looking at bees so a properly fitting pair of overalls is essential.

2.34: Beekeeping for all the family.

2.35: Providing protection for children need not be expensive.

2.36

2.37

2.36: I repair torn veils using a hot glue gun available from most hardware shops. The two edges of the torn gauze are held together and the glue is liberally applied. Hold together for a few seconds until the glue cools and hardens.

2.37: Beekeepers should keep a roll of duct tape in their tool kit in case their protective clothing rips when they are out in the apiary.

child, sew a small bee or other emblem onto the chest pocket of the overall, or even something totally non-bee related like a favourite cartoon character.

Gloves for children are available from some suppliers or alternatively kitchen gloves that are flocked inside can be used. Fit the gloves on the child over the overall sleeve. Buy some elastic and make a band to place over the glove so that there is a good seal preventing the possibility of any bees crawling inside. A further alternative is to buy a pair of child's gardening gloves from a discount store or hardware store and sew on a long gauntlet with an elasticised end.

I cannot stress enough that children must be supervised around bees at all times. This is particularly true if a child is given a hot smoker to manage and keep alight.

REPAIRING A TORN VEIL

There comes a time in the life of any protective clothing when the veil will tear and needs to be repaired. The best solution if the veil tears when the beekeeper is in the field well away from home is a little forethought, so always include some duct tape in your toolkit for emergency repairs both of the veil and the fabric of the clothing.

At home there are more options available to repair a torn veil and my favourite is a hot glue gun available from any hardware store. When hot glue cools it is fairly opaque and usually does not detract too much from the beekeeper's field of vision. Silicon sealant can also work, but this is harder to use due to its longer drying time and it is less opaque than hot glue after it has cooled.

SUMMARY

- The three items of equipment that you will need as a beginner are a hive tool, a smoker and a bee brush. These come in various designs and it is worth talking with other beekeepers about their preferences or visiting a beekeeping supplies store to look at what is available.
- There are three designs of protective clothing worn by beekeepers: the simple veil, a jacket with built-in veil, and a full overall with built-in veil. Gloves with long gauntlets are also needed to protect the hands from stings. These clothing items are critical for a beekeeper, so it is worth visiting a beekeeping supplies store to make sure that the clothing you buy is manufactured out of suitable fabric and is the correct size for you.

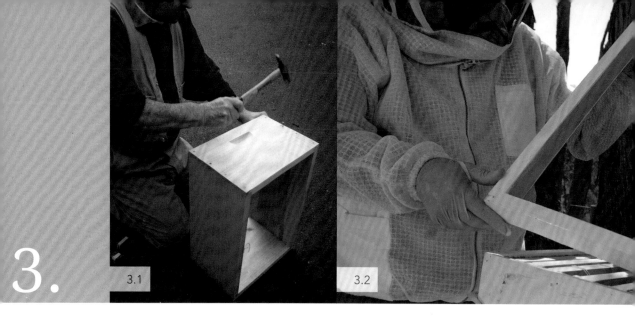

3.1

3.2

Constructing hives

Among many new beekeepers there is initially some confusion about the design and choice of hive ware available. The most common type of hive used in Australia and the one most readily available through suppliers is the Langstroth. Named after its American inventor, the Reverend Lorenzo Langstroth, the hives are essentially a series of oblong boxes, stacked on top of one another and held together with a base and lid. Langstroth designed his hive with removable frames, which was a breakthrough in beekeeping in the mid-1800s but is now a legal requirement for all beekeepers in Australia to allow for ease of inspection for disease.

While the boxes used in the Langstroth hives are identical, their use in the hive set-up determines the label by which they are referred to by beekeepers. The box placed on the hive base is called the brood box as this is where brood will be found, assuming, of course, that an excluder is used to restrict the movements of the queen and confine her to being able to lay eggs only in the bottom box. Any box above the excluder is called a super or honey super. It is exactly the same box in all respects, but defined and labelled by the role it plays in the hive set-up. The name super comes from the word 'superstructure', referring to the top-most parts of a building or structure.

If an excluder is not used to keep the queen in the bottom box, she will travel upward through the stack of available boxes and will lay eggs in the central frames in roughly an oval pattern. It is often for this reason that new beekeepers are encouraged to use excluders as, if nothing else, they know the queen and brood are in the bottom box and when it comes to taking honey they do not have to worry about avoiding brood frames

3.1: Making a hive.

3.2: A wire mesh top cover placed under the lid allows visitors to view the bees without wearing protective clothing.

throughout the honey supers. Of course, no theory is perfect and there are occasions when a smaller queen will slip through an excluder and go up into the top supers.

To further add to the confusion of the new beekeeper several sizes and depths of box are available from suppliers. Most hobby beekeepers use the full-size super and brood box. A description of each of the available sizes of boxes is given in the appendix.

CONSTRUCTING A HIVE

A complete beehive typically consists of:

- a base to rest the hive on
- a slotted bottom board, optional but increasingly used
- one or two brood boxes sitting on the base
- a queen excluder between the brood boxes and the supers
- one or two supers in which to store honey
- a hive mat used to stop worker bees building burr comb in the hive lid
- a lid to cover the hive that will protect the colony from predators and bad weather.

When I make up a hive in which to place a colony of bees I always place the queen in the bottom one or two boxes. The bottom of the hive is the natural place for the queen to lay eggs and for the nursery bees to raise brood. This arrangement allows the supers containing honey to be kept at the top of the hive where they can easily be inspected and the full frames of honey can be removed for extraction.

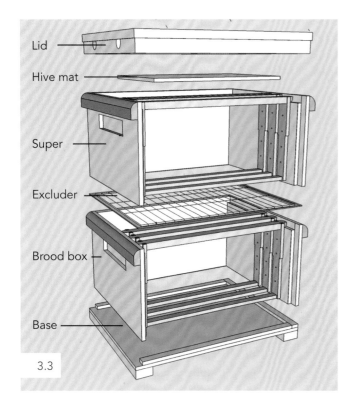

Lid

Hive mat

Super

Excluder

Brood box

Base

3.3

3.4

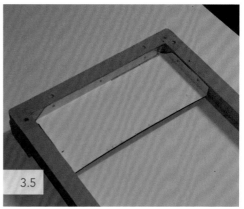

3.5

CONSTRUCTION OF THE PARTS OF A HIVE

THE BASE

The base of the hive protects the wooden hive parts from the damp by raising them above the ground. The base usually has built into it, around three of its top edges (two long sides and one short edge), a wooden riser that creates an entrance at the front of the base for the bees to enter and leave the hive through once a brood box is placed on top. Many commercially available bases have a riser attached on all four sides with a small entrance cut into the front riser for the bees to enter and exit the hive. In this type of base the width of the hive entrance is fixed and although it can be narrowed further by the beekeeper it cannot be widened should this be needed for extra ventilation. Risers can be obtained in thicknesses of between 10 millimetres and 50 millimetres depending on how much space the beekeeper wants below the brood box.

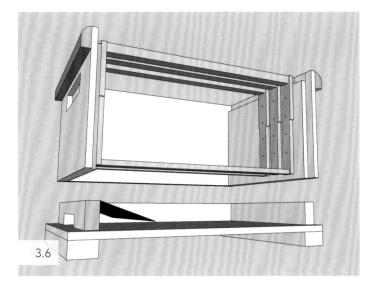

3.6

3.3: Schematic of a hive.

3.4: Hive bases come in many designs. This is a standard base with a solid bottom. The entrance size has been reduced although some designs leave the full front of the base open for returning foragers to enter and leave.

3.5: This design of base utilises a longer entrance way so that guard bees can better protect the colony against robber bees from other colonies.

3.6: A longer entrance way minimises the chances of robber bees or wasps entering the hive and stealing honey.

The flat bottom board or floor of the base is usually made of one of the following materials:

- wood
- plywood
- Weathertex
- galvanised steel
- moulded plastic.

The flat bottom board usually rests on a set of cleats made of either hardwood or treated pine. In plastic versions the cleats may be moulded as part of the hive base. Base cleats keep the bottom board a few centimetres off the ground and help keep the hive dry and free of wood rot. Some beekeepers nail the base to the bottom of the brood box. I don't recommend this practice as it makes life difficult when it comes to maintaining, cleaning or inspecting the base board, which in this case can only be done once frames are removed from the attached box. A freestanding base makes good sense, particularly with the advent of Small Hive Beetle and other pests. When maintenance or cleaning is required on the base it can be removed as a single unit rather than requiring removal of both the box and the attached base, which will thoroughly disrupt the hive's resident colony.

If *Varroa* spreads across Australia (see Chapter 16: Parasites of the Honey Bee), a slightly different design of base board will be required to help manage this pest. To assist with the control of *Varroa* mites overseas, the bottom boards are perforated by either using a stainless-steel mesh or by using plastic slats. The aim of this type of vented bottom board is to allow *Varroa* mites that have fallen off frames or bees to

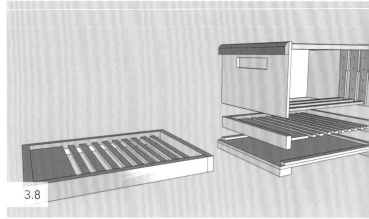

3.7: An American style base with a sloping landing strip attached to a separate stand.

3.8: A slotted bottom board distributes air flow more evenly throughout the hive. This minimises draft through the brood box near the entrance and encourages the queen to lay eggs in that area. During the colder winter months the design of the board enables the bees to keep the brood box warmer. The wide slat at the front allows guard bees to better protect the entrance to the hive against intruders.

drop through the hive onto the ground where they will die outside the host colony. This effectively helps to break the *Varroa* lifecycle and studies overseas have shown that the use of a vented bottom board effectively reduces *Varroa* mite counts by about 30 per cent. These modified bottom boards are discussed in more detail in Chapter 16: Parasites of the Honey Bee.

Another advantage of vented bottom boards, or aerated bottom boards as they are sometimes called, is that they allow a much greater flow of air through the hive. A US Department of Agriculture (USDA) study reported at the Apimondia beekeeping conference in Melbourne in 2007 stated that, with an increase in airflow past the colony, the queen lays far more eggs resulting in a much stronger and healthier colony. This result, however, was not substantiated in a more recent study of vented bottom boards conducted by Dr Doug Somerville of the NSW Department of Primary Industries. The hives that used vented bottom boards in Dr Somerville's study did, however, contain more honey than hives that did not use vented bottom boards.

Many hobby beekeepers ask me about the size of the landing provided at the front of the hive entrance. Although bees seem to adapt well to feral nests that have only a hole in a tree or a wall as an entrance, most hives have a landing at the front for the bees to rest on before entering or leaving the hive. During hot summer evenings it is common to see thousands of bees outside and hanging off the front of the hive in an effort to prevent the colony from overheating by having too many bees on the inside. It is my preference to use a larger landing with a depth of about 5 centimetres. This is in contrast to some of the smaller landings that have a depth of around 1 centimetre.

Professional apiarists prefer to use narrow or no landings since they can more easily pack hives close together during transport to new sites either for pollination or to follow the flowering patterns of gum trees. In practice, landing strips are commonly provided by using an extended length of hive base board; that is, the base board extends beyond the length of the bottom box. For the hobbyist with time, more complicated slanted boards, often seen in North American beekeeping books, can be constructed.

SLOTTED BOTTOM BOARD

Some beekeepers, myself included, place a slotted bottom board (not to be confused with a vented bottom board which is part of the base) between the base and the first brood box. The slotted bottom board is a separate unit from the base board, but once placed on top of this, it becomes a permanent part of the hive set-up. The use of a slotted bottom board offers the following advantages and disadvantages.

Advantages:
- The wide slat at the front of the slotted bottom board reduces the ease with which robber bees and wasps can enter the hive and fly up among brood frames. This is because the longer entranceway allows the guard bees greater time in which to fight off intruders.
- The use of slats distributes airflow throughout the entire brood box, not just at the front of the hive near the entrance. A queen is often reluctant to lay eggs near the entrance to the hive if there is a strong current of air coming in. The use of the slotted bottom boards encourages the queen to use the entire brood box for eggs.
- During winter, workers use their bodies to reduce the amount of cold air entering the hive and they are able to straddle their bodies across the gaps between the slats to reduce the flow of cold air. This effectively allows the workers to keep the hive warmer and they require less honey over the winter to survive.

Disadvantages:
- The use of a slotted bottom board does reduce the amount of air circulating through the hive during hot days. I believe, though, that the benefits outweigh this disadvantage and by the start of spring, hives that use a slotted bottom board are far stronger than hives that do not use them
- The use of a slotted bottom increases the overall cost of a hive.

BROOD BOXES AND SUPERS

As mentioned earlier, there is no physical difference between a brood box and a super box and the same box can be used for both purposes.

In Australia the wood used to make hives is typically 22 millimetres thick. This is thicker than the commonly available 19-millimetre-thick wood found in hardware and building supply stores. There is no doubt that the thicker 22-millimetre wood provides better insulation for the colony both during the hot summer and during the cold winter. The thicker wood also makes for a stronger and more durable hive. Many years ago when the imperial system of measurement was in use, 7/8-inch-thick wood was widely used in the construction industry and its use has continued to this day in the beekeeping industry with the conversion of 7/8-inch wood to 22-millimetre-thick wood for hive-making. Generally suppliers will offer hive boxes in either radiata or hoop pine.

CONSTRUCTION OF BROOD BOXES AND SUPERS

The corners of boxes are designed to fit together either using a rebated corner or by locked corners, often incorrectly called dovetail corners. For cost reasons rebated corners are the most frequently used corner for boxes, and when correctly assembled using glue and either nails or screws, make a very strong and durable corner.

Many beekeepers believe that the locked corner is stronger than a rebated corner. Although this may have been true in the past, modern glues are so strong that a correctly assembled, rebated corner is just as robust as a locked corner. Perhaps the main advantage of a box assembled using locked corners is that the box holds itself together firmly while the beekeeper is inserting nails or screws into the corners.

The disadvantages of a locked corner is that more of the wood's end grain is exposed, making the box more likely to rot along exposed corners. Interestingly, a lot of anecdotal evidence after the severe bush fires experienced in Victoria during recent years indicates that the hives made with locked corners suffered more from ember attack than those hives made with rebated corners. Another disadvantage of locked corners, and one to be aware of, is that the interlocking corner needs to be assembled soon after manufacture. Because the short sides and the long sides of boxes can be made with wood cut from different trees and dried to slightly different standards, leaving locked corner sides unassembled for long periods may result in one side contracting a greater amount than the other, making them difficult to interlock with each other. The result is that the box walls may not fit and can easily split if forced together. To be fair, though, the walls of boxes made with rebated ends can also shrink differently although this is usually much less of a problem when it comes to assembly.

3.9

3.10

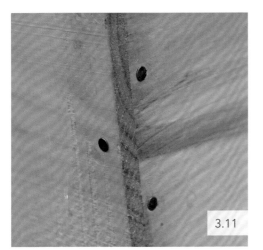

3.11

3.12

3.13

3.9: The slotted bottom board fits between the base and the brood box.

3.10: The interlocking corners of brood boxes or supers can be assembled with either 65mm galvanized nails or screws. If nails are used place five nails on one side of an edge and four nails on the other side. If 65mm long screws are used I only place them on one side of the edge due to their superior gripping power over nails.

3.11: This rebated box is joined together using 65mm nails. Modern glues provide excellent adhesion to wood corners. There is no disadvantage in strength in using a well-glued rebated box rather than an interlocking edge box.

3.12: Assembling a box with the aid of a clamp.

3.13: To stop the edge of the box splitting when you insert screws, this handy drill available from most hardware stores will both drill and countersink a hole at the same time. If you are using nails, holes also need to be drilled, but not countersunk, to stop the wood from splitting.

3.14: A full size super can weigh up to 35kg when full of honey. If the only way to hold the super is the grips cut into the wood these can be very awkward to pick up and carry. Adding cleats to a super makes lifting it easier, and they should be attached to all brood boxes and supers.

3.15: Bees usually propolise boxes and lids together, making them difficult to prise apart. Planing a bevel around the edge of the boxes and lid enables the hive tool to prise apart the box much more easily.

To assemble the boxes I use either 65-millimetre-long cement-coated nails or 65-millimetre screws. My preference is for galvanised screws along with a good wood glue as this gives the box optimum strength. A hive is roughly the same temperature inside throughout the year while temperatures outside the box, depending on its location, will vary a great deal. This will place a great deal of stress on the wooden corners of the box and the extra time and effort put into careful construction will pay dividends in the long term. If I use nails I put them into both sides of a corner for strength; that is, where the corner of both the long side and short side of the box meet. On the other hand, when I use screws I find these provide a sufficiently strong corner for the box even if they are inserted only along the corner of the short end of the box.

The preferred nails used to make boxes are usually cement- or glue-coated because of their superior adhesion to the wood when nailed in. 'Cement Coated' is the trade name for glue coated and the terms may be used interchangeably. As the nail is driven into the wood the friction causes the cement to melt and results in a much stronger attachment between the wood and the nail, creating a stronger corner.

I use five screws at each corner of a full-size or WSP box (named after William S. Pender, the inventor) and three screws or nails on an Ideal box. If I use nails to secure the corners, in addition to the five nails used on the short corner of the box, I also insert another four nails into the other corner on the long side of the box for added strength.

Most hive boxes come with pre-cut grips in the sides. The difficulty with pre-cut grips is that they are shallow and a full-sized super of honey is difficult to lift and hold using only your fingers. To overcome this I attach holding cleats made out of wood.

Holding cleats can be placed on either the long or short side of the box. My preference is for the short side. The use of holding cleats makes lifting the super a lot easier and you are much less likely to strain your back. The cleats are basically a strip of wood the same length as the side of the box on which they are to be fixed. They are around 4 centimetres wide and 3 centimetres thick and allow for a strong grip when lifting a heavy box.

The last part of box construction is painting. I only ever use exterior latex house paint, usually from the discard pile at the local hardware or paint shop. These discarded tins are usually well-known brands that have been rejected due to unsatisfactory colour mixing. I believe that two coats of a good-quality exterior house paint provides sufficient protection for my hives. Some beekeepers prefer to apply three coats of paint and sometimes even a primer. Other beekeepers use oil-based paint and there are those beekeepers who prefer to varnish their hives.

When buying paint do not skimp and buy cheaper brands as these frequently require three coats to provide a uniform, durable covering, whereas better quality paint will require only two coats to provide equal protection against the elements.

Hives are traditionally painted white as the idea is that this colour provides optimum reflection of the sun's rays and assists in keeping the colony cool during hot weather. These days, however, there is a move away from this practice, with many varying colours chosen by beekeepers. As a general rule, hives are not usually painted with black or dark paint, as this will absorb too much heat during the hot summer months. Keeping the colony cool during summer is more of a concern than keeping the same colony warm over a cold winter when a darker coloured paint would be beneficial.

Have fun with your hives if you have the time and inclination — they can be an attractive garden feature in their own right. When my children were small I painted birds and other characters on my hives and these not only delighted them but many of our visitors. If you feel you do not have an artistic bent, buy some children's stencils and use these. Be aware though, that particularly in the more northerly parts of Australia hive wood may be treated with copper naphthenate as a preservative. Copper naphthenate is poisonous and hives treated with this chemical should not be handled by children.

The debate about whether it is necessary to paint the inside of the box as well as the outside rages back and forth with both sides holding strong views. I no longer paint the inside of my hives because I do not believe that it is either necessary for the protection of the wood or desirable to have paint next to bees or honey.

On some of my hives I paint a distinctive pattern at the front for that hive's returning foragers to home in on. The reason I do this is that if there are several identically painted hives placed next to each other returning foragers may get disoriented and drift into a neighbouring hive instead of their own.

3.16: A creatively decorated hive.

3.17: Horses painted on side of brood box.

THE QUEEN EXCLUDER

Although not all beekeepers use queen excluders, there are many good arguments both for and against their use in a hive, as explained below. Excluders are placed between the brood box and the honey super to reduce the likelihood that the queen will leave the brood box and move up into the supers to lay eggs in the honeycomb.

Excluders are generally made of either galvanised steel or plastic, although some suppliers also offer a bamboo version. Any of these materials is suitable for use as an excluder. Plastic excluders are usually about half the price of metal excluders and many beekeepers choose them for this reason. The plastic in the excluder tends to become brittle over time, giving them a limited lifespan. Against this they have the real advantage that they provide no nooks and crannies for pests such as the Small Hive Beetle, *Aethina tumida*, to hide in. Further they do not conduct either heat or cold inside the hive. If you choose a plastic excluder check that it is of the type that has rounded grille bars, as plastic excluders cut out of one flat sheet of plastic may have sharp grille edges that will damage worker bees' delicate wings. Metal excluders are generally much easier to clean and will outlast a plastic excluder by many years.

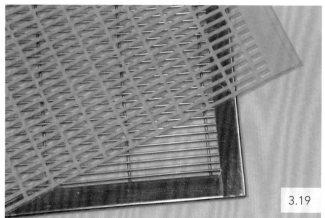

3.18: A great hive decorated by one of our Tasmanian friends.

3.19: Queen excluders are typically made of galvanised steel or plastic.

Beekeepers often buy excluders without giving any consideration to the gap width between the grille bars. It is a simple fact that both queens and worker bees vary in size. Most excluders have a gap width of between 4.1 millimetres and 4.4 millimetres. This is a large variation in grille size and some workers will be unable to get through a 4.1-millimetre gap while a few queens can get through a 4.4-millimetre gap. Before buying an excluder check if your bees are large or small and try to buy an appropriate excluder for their size, which is usually one with a gap size of 4.3 millimetres or slightly greater. To decide if your bees are large or small you may need to look at bees from other people's hives to get a better idea of the comparison.

SHOULD I USE A QUEEN EXCLUDER?

Many beekeepers choose to use queen excluders, although in my experience roughly 10 per cent or less use no excluders at all. I have, over the years, kept hives in which I used queen excluders and hives in which I did not use them. My personal preference is to place an excluder in a hive. Nevertheless, there are advantages and disadvantages in their use. These are summarised overleaf.

Advantages:

- The use of an excluder keeps the queen below the honey supers so that frames of honey can be removed without any concern about brood being present.
- Since the queen is unable to lay eggs in the super, the comb will remain white or light yellow without the unsightly dark wax caused by larvae skins, excreta and cocoons.
- Keeping the queen in a restricted area of the hive, the brood box, makes her easier to find when needed.
- Many workers will stay with the brood in the bottom brood box. This is an advantage when removing frames of honey since bees are more easily removed from honey frames than from frames containing brood.
- Pollen is usually stored next to the brood so there will be little pollen to discolour the honey in the super.

Disadvantages:

- Airflow through the hive may be reduced and this can cause overheating on very hot days. Brood are unable to survive heat as well as they can cold.
- Workers may experience difficulty getting through the grille of the excluder and be unable to store some of their honey in the super.
- Excluders are expensive and easily damaged.
- Some excluders may damage workers' wings as they struggle through the grille.
- If the queen inadvertently gets through the excluder into the super and lays eggs, when the larger drones emerge they may get trapped above the excluder in the super.
- Worker bees come in all sorts of sizes and a single gap width on a queen excluder may not be suitable for all the hives in your apiary. Queens also come in different sizes and the small virgin or newly mated queens may easily get through some excluders. The use of an excluder is for the sole benefit of the beekeeper and not the bee.
- The main issue is that the excluder constricts the size of the brood nest and reduces the capacity for growth.
- It may also encourage swarming.

3.20

THE HIVE MAT

The hive mat rests on top of the frames in the uppermost box under the lid and discourages the workers from building burr comb inside the lid. Burr comb is a nuisance to the beekeeper as it makes the lid difficult to remove. Also, once the comb has been broken it makes a mess in and around the top of the hive, creating sticky, unpleasant inspection conditions. The hive mat can be made of a variety of materials although I usually use linoleum purchased as off-cuts from a floor-covering store and then cut to size. Not every beekeeper uses hive mats and many people, particularly professional beekeepers, are content to let the bees make burr comb in the lid.

Hive mats are cut so there is a 1-centimetre gap between the edges of the mat and the inside walls of the hive. This is to allow air to circulate through the hive and out of the vents in the lid. Circulation is important in keeping the colony cool during the summer and also assists in the evaporation of condensation during the winter.

3.21

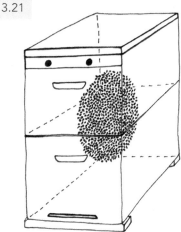

3.20: Plastic queen excluders are cheaper than metal excluders but are more likely to break during use.

3.21: If an excluder is not used in a hive, the queen will lay brood in an oval pattern at the centre of the hive, leaving the frames on the outside to store honey.

LIDS

The lid is a critical part of the hive since it protects colonies from heat, cold, inclement weather and from predators. Lids come in two designs: migratory lids and telescopic lids.

The migratory lid, named because of its use in migratory beekeeping, is the most common lid design used in Australia due to its lower cost, its utility when moving numbers of hives, and largely due to the fact that often it is the only design of lid that beekeeping supply stores sell.

A migratory lid is designed so that the sides of the lid are flush with the sides of the hive box. This allows many hives to be packed closely together when they are transported on a truck or a trailer. The disadvantage of a migratory lid is that it can allow rain to run from the top of the lid and possibly enter the hive either under the lid or where the boxes meet. This may happen when the lid is accidently knocked off centre and will cause the colony to be exposed to the elements and predators.

The telescopic lid, although less well known in Australia, is designed to overlap the top hive box and those boxes underneath by approximately 2 centimetres, allowing water to run harmlessly from the hive lid to the ground. Further, by overhanging the top box they afford better protection against rain beating in under the hive lid. It is a pity that these lids are not more popular among hobby beekeepers as they provide better rain protection to hives than do the more commonly available migratory lids.

The top cover of either a migratory or telescopic lid is usually galvanised or Colorbond steel. Increasingly Weathertex, a long-lasting, manufactured, resin-infused hardwood, is used, as this offers protection without conducting heat or cold into the hive. Ordinary wood can also be painted and used. I tend to use a mixture of Weathertex and Colorbond both for practical and aesthetic reasons. Both come ready painted and during the hot Australian summers a well-insulated Weathertex lid offers the bees better protection from the heat. Weathertex comes with a 25-year written guarantee against deterioration during bad weather and does not require painting.

Lids that have a wood or plywood top provide good weather protection to the colony provided the wood or plywood is well painted. Another part of the lid design that I am often asked about is the necessity for vent holes in the rim. Some suppliers supply lid rims with the holes for vents pre-drilled, others do not. It is usual to see two vent holes in the front of the lid rim and two at the back of the lid rim covered by either a metal or plastic vent, which prevents robber bees entering the hive. Although the primary use of vent holes is to allow the colony to breathe when the entrance is closed for moving, the use of vents also allows moist air to escape from the hive during winter or during rainy or humid conditions. If moist air condenses on the inner roof it soon forms drops of water which will run down the inside walls of the

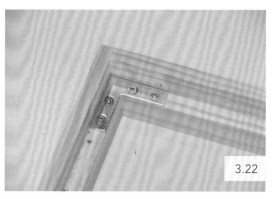

3.22

3.23

3.24

3.25

hive. This dampness will encourage the development of *Nosema*, Chalkbrood and other diseases in the colony.

Many migratory lids do not use air vents and the hive thus does not have the facility to allow damp air to escape. If the hive is kept over winter in dryer areas this may not be a problem, but I believe that a well-designed hive should have vents in its lid. Vent holes are often left out of lid rims during manufacture for cost reasons but they are easy enough to make with the correct drill bit. A hole diameter of between 15 millimetres and 25 millimetres should be used.

3.22: The steel corner angles used to join the telescopic sides of the wooden rim together also raise the rim a couple of millimetres above the super. This allows air to vent out of the super, keeping the hive dryer.

3.23: The telescopic lid overhangs the sides of the hive and provides better protection to the colony against wet weather. A disadvantage of the telescopic lid is that it is much more difficult to pack together hives for transport. Underneath the telescopic lid in this photograph is a top feeder as well as insulation.

3.24: The sides of a migratory lid are flush with the sides of the super, allowing more hives to be packed together for transport.

3.25: Placing a hive mat above the top set of frames under the lids discourages the bees laying burr comb in the lid. The hive mat is cut so that there is about a 1cm gap between the mat and the sides of the box so that air can still circulate into the lid.

It is worth noting that some bees will consistently propolise lid vents if provided, but I am happy to give them that choice. Many experienced apiarists believe that if the vent is attached on the outside of the lid the bees are less likely to propolise it, although none of my inside vents has ever been closed by my bees. Beware though, that spiders are able to make nests in vent holes if the grille is attached to the inside and can bite fingers when the lid is removed.

Depending on their design, telescopic lids use a different technique to allow for moist air to circulate out of the hive. In one common design the telescopic lid is raised about 2 or 3 millimetres above the top edge of the super. This small gap provides ample ventilation for a moist hive to vent damp air without being wide enough for the bees to exit or pests to enter the hive. The gap also prevents workers propolising the lid to the super, making it easier to remove.

In many respects this type of lid seems an improvement on the commonly used migratory lid and I believe that its use will grow in popularity in the coming years. Of course, the inventive practical beekeeper is already making design improvements to the standard migratory lid but the majority of beekeepers will buy their hive ware in a kit form from a beekeeping supply store. The popularity of new lid designs will be influenced by the willingness of suppliers to stock and explain their use and, of course, by beekeepers willing to pay a premium to try something different.

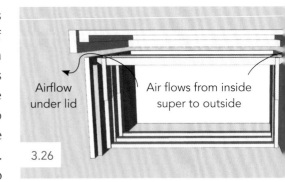

Airflow under lid

Air flows from inside super to outside

3.26

3.27

3.28

3.26: Airflow out of a hive with a telescopic lid.

3.27: It is a legal requirement across Australia that all hives be marked with the registration number of the owner.

3.28: Plastic hives are usually well designed and come in different colours for ease of identification.

With the standard North American design for the telescopic lid, a crown board is needed to provide ventilation since this design does not include a 2- or 3-millimetre gap above the top edge of the super. Also, with the standard North American design a top board is needed since the bees are able to propolise the lid to the super and a hive tool is unable to reach under the lid to prise it open.

In Australia there is a legal requirement in each state and territory for each hive to be indelibly marked with the registration number of the owner. Most beekeepers paint their number on hives although some brand their number as well as marking their frames, as a further safeguard against theft.

PLASTIC HIVES

Plastic hives are relatively new to the beekeeping scene in Australia. There are now several types available that take eight or ten full-depth frames. Whether their popularity will grow to the point where they replace a high percentage of wooden hives remains to be seen. These hives are usually ready assembled, are convenient to use, will never need painting and will not rot. While they may be durable and convenient I have a philosophical objection to housing my bees in plastic. I may be old-fashioned and inflexible but that is choice I have made for my own apiary. I must now own up to the fact that I have minimal personal experience with the plastic hives available and much of what I have discovered has been from anecdotal evidence provided by the customers of my business. Some like them and, of course, others do not — just like everything else in life.

My only actual experience with a plastic hive is with one that was sent to me as a demonstration model. I did not house any bees in it but stored it outside until the sender arranged to collect it from me. We had one of Melbourne's 42°C days and the sides of the box buckled under the heat and the frames fell in. Interestingly the box returned to its normal shape during the evening so I was unable to photograph the distortion in the hive the next day.

OTHER PARTS OF A HIVE

ADDITIONAL WEATHER PROTECTION ABOVE THE LID

I believe that it is beneficial to the bees in the hive to provide additional protection against bad weather above the lid. A piece of plywood or cement board can be used together with some pine to keep the additional cover raised off the existing hive lid. In addition to the above, during the very hot days of summer, to allow greater air circulation through the hive I will often slightly open the existing hive lid, moving the top cover further over to cover the gap in case of rain.

3.29: Hot sun baking down on a metal hive lid can cause the colony to overheat. I cover my hives with this simple lid made of scrap plywood to provide additional protection against the hot Australian sun during the summer months.

3.30: On hot days a lot of heat can enter the hive through the roof. To help minimise this, I sometimes place an Ideal box containing dry gum leaves between the top super and the lid. To stop the gum leaves falling into the hive I tack fly wire underneath the Ideal box. The box can also be filled with crumpled paper or other suitable insulating material.

3.31: This lid rim has a layer of fly wire attached to it so that the beekeeper can remove the lid without the bees coming out to sting. This is a great idea for people who want to show visitors or children their bees without getting them dressed in protective clothing. These screened lids are also great for moving sealed colonies since they allow air into the hive while stopping bees from escaping.

3.32: Instead of a wire mesh top cover, a cover made of Perspex can be placed under the lid to allow ease of viewing.

INSULATION UNDER LID

This idea originated with the Warré hive. I sometimes provide additional insulation under the lid by filling an Ideal box with dry gum leaves or crumpled newspaper and placing it between the top super and the lid. The leaves are prevented from falling into the super by covering the bottom of the Ideal with wire netting of the type used in flywire screen doors.

SUMMARY

A hive consists of:
- a base
- one or more brood boxes
- a queen excluder
- one or more supers
- a hive mat
- a lid.

Although they are called by different names, both a brood box and a super are identical. The use to which they are put in the hive set-up to either house the queen and brood or to store honey determines what they are called. Hives usually come in eight- or ten-frame sizes; they can hold either eight or ten frames per box. The height of the box can vary and the choice is usually made on the basis of weight since a full-size ten-frame super of capped honey can weigh as much as 40 kilograms. Ideal or WSP size supers are gaining in popularity since they are much lighter to lift when full of honey. Many beekeepers are choosing to use hives made of plastic rather than wood as they offer convenience, low maintenance and durability.

4.1

4.2

4.

Frames and how to build them

FRAMES

The frame is a key part of the hive and its design, assembly and ongoing maintenance is more critical than that of any other part of the hive. The frame is made up of four pieces of wood:

- a top bar
- two side bars
- a bottom bar.

Because frames need to hold and support both brood comb and honeycomb, assembled frames need to provide support for the sheets of beeswax or plastic foundation that are inserted into them. To provide support to beeswax foundation sheets, a full-size frame usually includes four taut wires running horizontally between the side bars. Since the wire is taut, eyelets need to be inserted into the pre-drilled holes in the side bars to stop the wire cutting into the wood when the frame is holding a fully capped load of comb and honey. It is possible to use staples to do the same thing and several punched in to hold the wire on the side of the frame will take the strain off the wood.

The wooden side bars, the top bar and the bottom bar are usually held together using cement-coated nails with a size of approximately 1.4 millimetres x 35 millimetres.

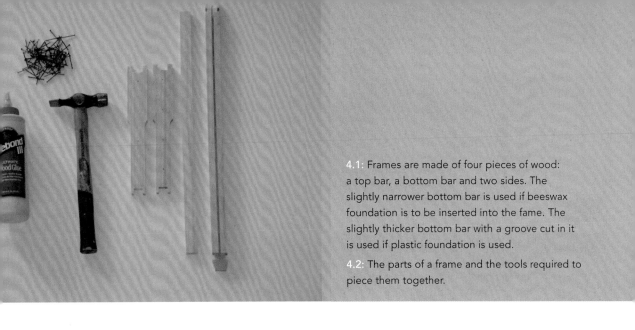

Two nails are placed on either side of the top bar and hold it to the side bar. The bottom bar receives one nail on either side. Finally, a small tack is inserted on the top part of the side bar where it joins the top bar.

The reason for the extra nails on either side of the top bar is that it is the top bar of the frame that takes all of the strain. It will hold the weight of a full frame of comb and also be repeatedly levered out of the hive with a hive tool by the beekeeper. Again, staples can be used and some beekeepers find it easier to use them in preference to nails. Many commercially available pre-assembled frames are stapled. Using a good-quality glue such as PVA or Titebond III prior to nailing or stapling also helps to keep the frame together and add to its useful life.

FRAME WIRING BOARD

Frames can be wired by hand with a minimum of tools although I find that a frame wiring board as shown in Visual 4.12 makes the wiring much faster and much simpler than other methods.

FRAME-MAKING JIG

Most hobby beekeepers with a small number of hives need only assemble frames one at a time. For beekeepers with a larger number of hives who need to assemble many frames at the same time a frame-making jig will prove invaluable. A frame-making jig for full-sized frames can be made out of a modified WSP box. The use of a jig will significantly reduce not only the amount of time it takes to assemble frames, but will also make their assembly a much easier task.

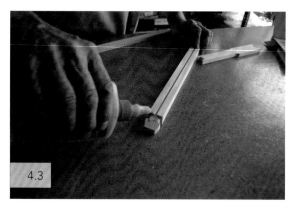

4.3

4.4

4.5

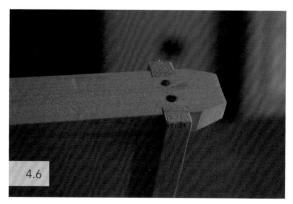

4.6

4.7

4.8

4.3: Step 1: Before joining the four pieces of wood together, first place a small amount of glue at each join.

4.4: Step 2: Join the four pieces of wood together ready for nailing. You may need to gently tap the corners to get them to fit together snugly.

4.5: Step 3: Next nail the top and bottom bar to the sides.

4.6: Step 4: Since the joins on the top bar have the most stress on them, use two nails through the top and another tack through the side for added strength.

4.7: Step 5: Placing a tack through the side of the top bar for added strength.

4.8: Step 6: The bottom bar needs only one nail through each end since there is not much stress placed on these joins.

4.9: Step 7: Two tacks are inserted into the sides of a side bar to attach the frame wire.

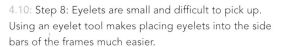

4.10: Step 8: Eyelets are small and difficult to pick up. Using an eyelet tool makes placing eyelets into the side bars of the frames much easier.

4.11: Step 9: Using an eyelet tool, insert eyelets into the side bars of the frames.

4.12: Step 10: One of the tricky parts of attaching wire to frames is getting the wire tight. I use a frame wiring board to do this.

4.13: Step 11: Threading the wire through the eyelets using the round door stops attached to the frame wiring board makes pulling the wire easier.

4.14: Step 12: The tricky part of attaching wire to the frame is getting the wire tight. To do this, use the lever to push the side of the frame inwards.

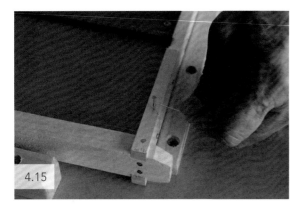

4.15

4.16

4.17

4.18

4.15: Step 13: Tie off the wire onto the two tacks, release the lever, and the side bar returning to its correct shape will tighten the wire.

4.16: Step 14: Once the wire has been tied off, cut the ends and hammer in the tacks until they are flush with the side bars.

4.17: Step 15: The wire has been tied off on the tacks and, once the lever has been released, the wire is taut.

4.18: Some beekeepers attach the wire vertically.

4.19

4.19: If you make lots of frames it is worth building a frame-making jig. Here I have used a modified full-size box as the jig. To make the jig cut a small amount of wood from the bottom of the box so that the frames rest on the floor when placed inside the jig. I have attached a Velcro strap across the top of the box to hold the frames in when the box is turned upside down to hammer nails into the bottom bars.

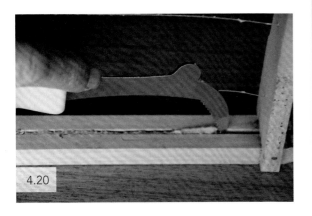

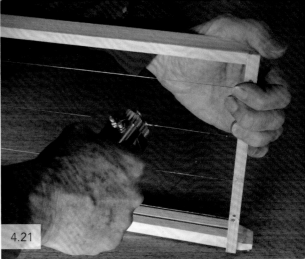

CLEANING USED FRAMES

Frames that contain black comb or those that have been damaged by Wax Moth need not be discarded but can be cleaned and reused. First remove the entire wax comb and then clean out the groove in the top bar. Next remove all of the old wax from the wire so that the new foundation will be easier to attach. If necessary tighten the frame wire using a crimper.

4.20: Re-using old frames often requires that the groove in the top bar be cleaned. A cleaning tool is used to remove the wax from the groove.

4.21: Before re-using an old frame the frame wire sometimes needs tightening. Here I am using a crimper to tighten the wire.

ATTACHING FOUNDATION TO THE FRAME

Attaching the wax foundation inside the frame is one of the two tricky parts of making a frame; the other trick is getting the supporting wire taut. The best method of attaching wax foundation is to use a car battery charger to heat the wire so that the wax foundation melts into it. A 12-volt car battery charger, if used alone with stainless steel frame wire, would generate too much heat, may melt the wire and will damage wax sheets in an instant. It is better to use a resistor in series with one of the charger leads to reduce the current. I use a 2-ohm, 50-watt resistor to embed foundation onto stainless steel frame wire. This size resistor is unsuitable for use with galvanised or tinned wire, as the wire will not heat sufficiently for the foundation to melt into it. It is worth noting that many expensive car battery chargers cannot be used for embedding foundation as they often contain sensors that will shut off the charger if a car battery is not detected.

Stainless steel is the preferred material for frame wire since it does not corrode when it comes in contact with the natural acids contained in honey. Galvanised or tinned wire was previously used to make frames, but these two materials are no longer popular with beekeepers due to their corrosive properties. An important dimension

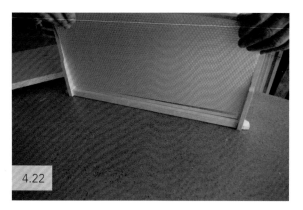

4.22

4.23

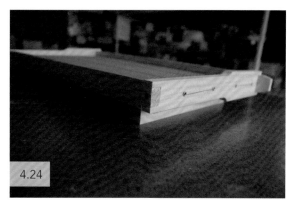

4.24

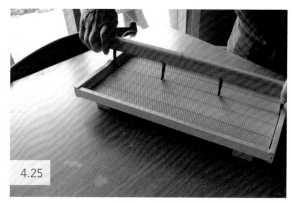

4.25

4.26

4.22: Attaching and inserting wax foundation. Step 1: Place a sheet of wax foundation into the frame making sure that the wax fits snugly into the slot in the top bar.

4.23: Step 2: Rest the foundation on a frame board with the wire resting on top of the foundation.

4.24: Step 3:The frame-making board needs to be sufficiently deep that the sides of the frame do not touch the table. This will ensure that the wire is resting on the foundation.

4.25: Step 4: Use an embedding tool connected to a car battery charger to attach the wax foundation to the wire. Connect a resistor in series with the charger otherwise too much current will flow and the wire will melt.

4.26: Step 5: Alternatively, use an embedding tool to press the wire down into soft wax. This needs to be done on a warm day or indoors when the wax foundation is soft and wire can easily be pressed into it. The use of an embedder is not a hard technique to learn but it is not as effective as using a battery charger to attach wire to foundation.

when designing frames is the bee space. The bee space is about 10 millimetres wide and is the minimum gap that two bees can pass through at the same time when working on opposite comb. If the distance between the two combs is greater than 10 millimetres the worker bees will often fill it up with burr or bridging comb making it difficult and messy for the beekeeper to remove frames.

Although sheets of wax foundation come pre-cut for full size or other types of frames, I often use strips of wax or broken sheets of wax in my frames. The bees do not mind and soon build out the foundation to fill the frame.

4.27: Instead of using a full sheet of foundation I often insert thin strips of foundation to provide a starter for the bees to build their own honeycomb. Using a thin starter strip is a good idea if you plan to eat the honeycomb since the comb will be much thinner and more delicate to eat than manufactured foundation.

4.28: A thin strip of foundation being built down by the bees.

4.29: A thin strip of foundation completely built out by the bees.

4.30: Even broken pieces of foundation can be used to make up frames.

USING WAX OR PLASTIC FOUNDATION

There is a growing trend for professional beekeepers to use plastic foundation in frames rather than foundation made of beeswax. Although plastic foundation is stronger and less likely to break during honey extraction, bees will generally only draw or build comb on pre-waxed plastic foundation when there is a honey flow on or if they are being fed sugar. This is because bees prefer to build comb on natural beeswax and will only draw comb on plastic as a last resort. Once they have initially drawn out the plastic foundation with comb, however, they will keep using this comb without any ongoing resistance.

If you do decide to use plastic foundation the surface of the plastic sheet will need to be prepared by coating it with a thin layer of beeswax, otherwise the bees will not accept it. To coat the surface I first melt some beeswax in an old skillet or wide pan. Then, using a small paint roller, I dip the roller into the melted wax and run it over the foundation sheet, first on one side and then on the other.

4.31

USING PLASTIC SIDES FOR FRAMES

The majority of people use wooden frames but there are many suppliers of plastic frames. These are convenient to use and come in a variety of colours and designs to suit the beekeeper.

4.31: Before placing plastic foundation into a hive it needs to be coated with wax. I melt the wax in a skillet and apply the wax using a paint roller.

4.32: Flexing plastic foundation into a wooden frame. Frames made to hold plastic foundation have a groove cut into both the top bar and the bottom bar.

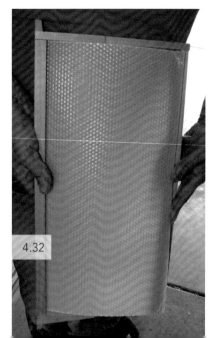

4.32

4.33/4.34/4.35/4.36: Plastic frames are sometimes used instead of wooden frames and come in many different designs and colours.

SUMMARY

- Frames are used in the hive for the workers to build brood comb and honeycomb on. Most frames are made of wood while the foundation that they hold is usually made of pure beeswax. Frames are more difficult to make than the beehive and the beginner will need to learn the techniques used to both tighten the frame wire as well as attach the beeswax foundation.

- There is a growing trend for beekeepers to use both frames and foundation made out of plastic. This is due to the superior durability and strength of plastic and, to some extent, the fact that frames made of wood can be time-consuming to wire and embed with foundation. In my experience, however, bees prefer foundation made of beeswax over foundation made of plastic.

Your first bees

In an ideal world every new beekeeper should either pair up with an experienced beekeeper or at least join a local beekeeping club. Beekeepers in clubs are generally friendly, helpful and eager to offer advice to new beekeepers. They are usually passionate about their hobby and proud of their hives. Even if you do not yet have your own hive, you may receive an invitation from a club member to inspect their hives, opening the supers and brood boxes under their guidance. Seize every opportunity and absorb the information given — it may contradict the book you have just finished reading but it will help you sort through any conflicting advice and information as you increase your own knowledge and practical experience. Clubs offer a lot to a new beekeeper as many have resources that can be borrowed including books, DVDs and extractors. Introductory workshops and courses on other aspects of beekeeping may be offered and specialist guest speakers are frequently invited to meetings. It is the easy access to a wide pool of varying experience that makes a club so valuable for the beginner.

Even knowledgeable experienced beekeepers need specialised advice at times and it is common to see and hear them sharing information and offering assistance to each other at a bee club. If you do not have a club in your part of the country there are apiarist associations throughout Australia and these can often be a good starting point for information. Some beginners prefer to pay an experienced beekeeper to offer them some initial assistance and if this is within your budget it might be an alternative to a club if there is not one near to your local area.

5.1: Honey bee pollinating an almond flower in north-west Victoria.

Bees lead complex lives and a lot about them remains unknown. There are endless discussions even amongst professional beekeepers about the best way to manage their colonies under different conditions — and what works in colder Tasmania may not work in Queensland where tropical conditions make beekeeping very different.

OBTAINING YOUR FIRST BEES

If you live near a large city that has a beekeeping club this will often be your starting point, as club members will provide advice and guidance on obtaining your first colony of bees. As a new beekeeper it is often more convenient to buy an established hive and to move this onto your property than it is to build a new hive and to arrange for either a swarm or nucleus of a queen and bees to be placed in it. Existing hives from a reputable supplier usually come with a guarantee against disease and the seller will often help to move and relocate the hive, advising you of a suitable aspect and location for bees on your property. Another advantage of buying an established hive is that you can ask the owner about the age of the queen, temperament of the bees, frequency of swarming and how much honey they produce. The answer to this latter question could be different once the colony is relocated to your area, which may not offer the same conditions for the bees.

On the other hand if you buy a hive from, say, an online site without any prior knowledge of beekeeping you will be unable to ask these questions and may be taken advantage of by an unscrupulous seller and sold a diseased hive, one riddled with Wax Moth or one so neglected that you may need to replace the frames and some of the hive ware. You will not have the skills to recognise the problems inside the hive and may be buying an expensive mistake.

Numerous people, including myself, have purchased an existing hive and then started learning about the bees inside it. With the benefit of hindsight I do not recommend that you do this and feel that the common sense approach is to learn about bees before you undertake to keep them. It is unreasonable to expect a more experienced beekeeper to give you valuable time sorting out any problems when you yourself could not make the effort or take the time to acquire some initial knowledge. You have a responsibility to register and to manage your hives in such a way that they are not a nuisance to your neighbours. You cannot do this without the benefit of some prior understanding of the management of your bees. Even if you buy a hive from an established beekeeper you should ideally ask someone more experienced to inspect the colony inside the hive before accepting it to ensure that all is well with the bees. Of course this is not always practical, particularly if you know of no other beekeepers or live in a more isolated rural area. At the very least you should satisfy yourself that the hive has a steady stream of bees coming and going from the hive entrance. This may well mean one trip to the supplier to look at the hive during daylight and then another visit to collect the hive either in the evening or early morning. Do not buy hives that look obviously neglected, rotting or if the seller mentions they were Dad's but have been sitting in the paddock untouched for a few years. These types of hives require a lot of work to sort out and are not really the introduction a beginner should look for into the world of beekeeping.

WHAT KIND OF BEES SHOULD I KEEP?

The three races of honey bee sold by queen rearers in Australia are the Italian, the Caucasian and the Carniolan bee. Beginner beekeepers often ask which is the best race of bees to keep and I have listed below the different characteristics of the three races.

Apis mellifera ligustica — Italian bees

- A slightly smaller bee than other races, the main distinguishing feature being the yellowy and straw-coloured or light brown stripes across its abdomen. Italian bees tend to cling more tightly to the comb when the frame is shaken and are regarded as being good honey producers, less aggressive and easier to handle than other races. Italian bees build up numbers later in the spring than other races although they soon make up these numbers as spring progresses.

Apis mellifera caucasica — Caucasians

- This race of bees originated in the Caucasus region north of the Black Sea and come in two types: the Mountain Grey strain from higher elevations and a yellowish strain from the lower elevations. The strain from the lower elevations tends to be more aggressive and stings freely while the Mountain Grey strain is gentle, industrious and survives winters well. They are good honey collectors although they tend to propolise hives more than other races.

Apis mellifera carnica — Carniolans
- This bee closely resembles the German black bee both in appearance and habits. Carnies, as Carniolan bees are affectionately called, are docile although they have a reputation for swarming. They are, however, good collectors of honey.

In spite of the above discussion on the different races of queen bees sold in Australia, most bees found here are a mixture of Italian, Caucasian, Carniolan and dark German races. This is because even if the virgin queen is genetically pure, there is less control over the genetics of the drones she mates with, resulting in workers that are hybrids or cross-breeds.

SHOULD I BUY AN ESTABLISHED HIVE OR CATCH A SWARM?

The advantage of buying an established colony is that you obtain a ready-made hive with an established colony of bees that can be placed immediately on your property without too much work. For the new beekeeper, the possibility of an established colony swarming during early spring is very real and may be a reason not to buy an established colony at this time. Although swarming is a natural part of the bee's lifecycle it takes experience to manage. If you are in the fortunate position of having an experienced beekeeper to offer assistance with your first hive swarming this will provide an excellent learning experience and allow you to hopefully approach the following spring swarming with some confidence.

An alternative method of obtaining bees is to first assemble your own hive and then catch a swarm to live inside it. If you are a member of a beekeeping club or know a local beekeeper they will often obtain a swarm for you. They may invite you to go with them to collect the swarm or alternatively you will provide an empty hive and the beekeeper will house the swarm inside it. Sometimes there is a small fee involved and in my own club there is also a prerequisite that a new beekeeper must have completed an introductory beekeeping course prior to being given a swarm. The advantage of a swarm is that it does not usually contain that many bees and these are less likely to swarm again during their first year. This allows time for the new beekeeper to gain knowledge and practical experience during the first year, ready for the swarming season the following year.

5.2/5.3/5.4: Swarms often land in easy to access places such as on the side of a shed, on tree stumps near the ground or some swarms make a permanent nest under the eaves of houses. This nest has lived under the eaves of this house near Lake Boga, Victoria, for the last two years.

5.5: The same permanent nest near Lake Boga when the bees moved away temporarily after part of the comb collapsed during very hot weather. The well-developed brood and honeycomb can be clearly seen.

5.6: Swarms often make a home in possum boxes. This box containing a feral nest is ready to be taken down from the tree and the colony rehoused.

5.7: Many swarms land in trees until they find a new home to move to.

One further option available is to buy frames of brood and adult bees. Many beekeepers will sell you four or more frames of brood and bees together with a viable queen to place in your hive to start a new colony. These may be placed in your own hive box or sold to you in a four- or five-frame nucleus box. If you obtain your colony in a nucleus box it is worth leaving it alone for a few weeks to allow the bees to orientate to your property.

SWARMS AND SWARM CATCHING

Most new or potential beekeepers, unsure of their skill level, are often reluctant to attempt to catch a swarm. If, however, a swarm is located in an easy-to-reach

5.4

5.6

5.7

place, say on a low overhanging branch or on a fence post, then the task of collection is usually quite straightforward. Many swarm catchers become star attractions while removing swarms from public areas and can draw quite a crowd. Although members of the public should be kept at a safe distance it is a fascinating sight for them to see a mass of bees march into a box to join the rest of a collected swarm and queen.

Swarms are generally very docile and even tempered since they are not protecting a colony containing brood or food; this is different to the aggressive temperament of nests of bees that are disturbed and immediately go on the attack to defend an established colony. Dressed in protective clothing, the swarm catcher only needs to brush or shake the swarm into a nucleus, brood box, cardboard box or bucket to collect them. Once the queen is in the chosen container the remaining bees will soon follow. It is worth mentioning that if a swarm has been in a location for several days and has started drawing comb then it is possible that they will become agitated when disturbed.

Swarms are readily available in the swarming season, usually between early September and late December.

Many swarms are located in difficult to reach places such as on high branches in trees or difficult to access parts of buildings. If the new swarm is located deep in shrubbery, a pair of secateurs will be a useful addition to the swarm collector's toolkit as these will efficiently remove obstructing smaller branches. A small saw is also useful to cut larger,

5.8

5.8: A bucket attached to a long tent pole is sufficient to bring down many swarms from high up in trees or on the sides of buildings.

thicker branches so that the branch and any attached swarm will fall or can be removed to a level more accessible to the swarm collector. If the swarm is located high in a tree, say up to about 5 metres, a plastic bucket attached to an extendable tent pole is useful. The bucket is raised until it is couple of centimetres below the swarm, then with an upward thrust the bucket is knocked into the branch directly where the swarm is located. The majority of the swarm will fall into the bucket, which is quickly lowered and the bees poured into a nucleus or similar box. The bucket is then raised again and the process repeated until most of the bees have been collected. Again, if the queen has been caught and is inside the holding box, the remaining bees will soon join her.

In most cases, even though swarms are not usually aggressive, the swarm catcher needs to use some ingenuity to find ways to access the colony and remove them. I would not allow the thought of possible difficulties to put you off trying to catch swarms that are in easily accessible locations. Catching, handling and managing swarms is a rewarding and educational experience and a useful and informative part of the beekeeper's art. In my experience, beekeepers who are prepared to collect swarms become better and more confident beekeepers as a result. Most feel a real sense of achievement plus they make an obvious contribution to the safety of their local community and save the swarm from possible extermination by other less 'bee-friendly' people or agencies.

An alternative way to attract a swarm, should one happen to be passing close by or if the bees are in a difficult to reach place, is to place a bait hive near the swarm's location and hope that scout bees will find the box, determine that it is a good home and return to the swarm to let them know of the bait hive's location. To improve the chances of this you can insert a vial of Nasonov scent inside the bait hive to attract the

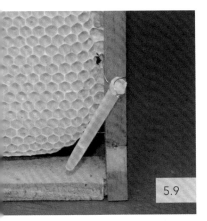  

5.9   5.10   5.11

scouts. Nasonov scent is the pheromone from the Nasonov gland, a small gland located above the stinger of the worker bees. Nasonov scent is used by bees to mark their colony and is used as a homing signal so that returning foraging and scout bees can more easily find their own colony. A few beekeeping supply shops have a chemical Nasonov for sale and this can be inserted into a bait hive as a swarm attractant. If you are attempting to catch swarms from your own apiary or hive it is generally recommended that bait hives be placed approximately 100 to 250 metres away from the hive that is about to swarm.

5.9: The Nasonov pheromone is the bees' natural homing signal. By placing a vial of Nasonov pheromone in an empty hive, preferably together with some dark foundation, it is often possible to catch your own swarm.

5.10: Nucleus boxes are frequently used by beekeepers to catch swarms and move them to a permanent location. The nucleus entrance can be sealed with a rotating coloured disc.

5.11: The difficulty when storing a swarm inside a nucleus and moving it long distances is that the colony may not have sufficient fresh air to breathe. At the side of my nucleus boxes I cut large holes and cover them with stainless steel meshing. This allows plenty of fresh air to enter the nucleus if I seal it for long periods

If a member of the public telephones to ask you to catch a swarm on their property you should consider charging a fee for this service. A lot of your valuable time and petrol will be used collecting swarms and many people who ask for swarms to be removed really do not appreciate the effort involved. Indeed, on occasion you may have to return to the property more than once.

After the captured swarm of bees are in residence in their new home, you could add a frame of open or capped brood comb from another hive since the presence of this will make it much less likely that the swarm will leave. Another possibility if the swarm is to be housed in a nucleus box for a while is to consider the use of a nucleus entrance disc. This is a rotating, coloured disc-shaped piece of plastic that is essentially divided into four different entrances that can be selected by the beekeeper. One of these has a queen excluder, which enables the bees to come and go from the nucleus as the swarm is establishing itself, but will prevent the queen from leaving. Effectively, if she stays, there is an excellent chance that the bees will do likewise and the captured swarm will establish into a viable colony.

If the captured swarm is weak, either merge it with an existing colony or add some frames of brood to build up the number of workers.

Generally, as the spring swarming season approaches, the experienced beekeeper prepares for this by having a number of items ready and waiting should the opportunity arise to collect a swarm. These may include a nucleus box with undrawn or drawn frames, a strong cardboard box, gardening secateurs, a small saw, an Emlock fastener (named after its inventor, Mr E.M. Lock of Victoria), an old sheet, a sharp knife and a water spray bottle to calm the bees. A large, strong, flat piece of cardboard is useful to have if the bees are to be knocked from a branch and into a container or box. The cardboard can be placed under the area of the swarm before the collection commences. Any bees that fall to the ground will not get tangled in long grass or other debris (on the site) and can more easily access the nucleus or other collecting box, which should be placed at one edge of the cardboard.

TRANSFERRING AN ESTABLISHED COLONY

A more challenging task is to remove an established feral colony with brood and place it into a hive box. This is a messy task and more time consuming than catching a swarm. It is often referred to by beekeepers as 'a cut out', as the beekeeper will literally cut the comb from the feral nest to attach to some ready-prepared empty frames. Many feral swarms are located in possum boxes, walls of buildings, inside roof spaces or other difficult to reach places. The swarm collector needs to decide if it is practical to remove the colony, leave them there or kill them. If you are planning to move a feral colony from a possum box, bird box or elsewhere to a hive, make sure that you do this during the spring or summer when the colony has the best chances of surviving the move. Moving a feral nest during the autumn or winter is likely to result in the death of the colony.

I will go through an example of a feral colony that has made its home in a possum box since this is a common situation, particularly in my local area.

5.12

5.13

5.14

5.15

5.16

5.17

5.18

5.12: Rehousing a swarm from a possum box. Step 1: Place the possum box on a plastic sheet and remove the lid and one or two of the sides. This will allow you to gain access to the comb inside the box.

5.13: Step 2: Once the comb is easy to reach, cut sheets of brood comb out of the possum box. A sharp knife with a long blade is best for this.

5.14: Step 3: Next attach the sheets of brood comb to an empty frame using string. Concentrate on getting as much brood comb out of the possum box as possible.

5.15: Step 4: Each sheet of comb is attached tidily to the frame.

5.16: Step 5: Once you have a frame full of brood comb insert it into an empty full size hive. Repeat this until the hive is full of frames containing brood. When you have collected as much brood comb as possible, attach honeycomb to some empty frames and place them into the hive at the edges of the box.

5.17: Step 6: The site is now tidied up and any remaining bees are brushed into the hive.

5.18: Step 7: The rehoused swarm is now ready to be moved to its permanent location once the Emlock fastener has been attached to keep the hive in one piece and the entrance sealed.

Initially the possum box will need to be removed from the tree and brought to ground level. There are various ways of doing this which I will not deal with here, although moving the colony from the possum box to the new hive needs to be done during the day when you can see clearly what is happening and are not rushed by oncoming dusk.

Prepare your working surface with a large piece of plastic, which can be placed on either the ground or an outside table, and will make the clean-up of the resulting mess a lot easier. Once the possum box is on the ground or the table, carefully remove one or two of the walls of the box, taking care to minimise damage to the comb and not to overly excite the bees. Tools such as a screwdriver, hammer and saw will be required to take apart the box. I find a strong screwdriver more useful than a hive tool as it is easier to hammer the handle-end to force open joints in the wooden box using the screwdriver end. The possum box should be taken apart sufficiently that the hanging comb can be cut from the roof in slabs.

Before you start cutting out brood comb, depending on how the nest has been assembled by the bees, shake or brush as many bees as possible into a hive, making every effort to get the queen into the hive as well. Inside the hive there should be some frames of drawn comb so the queen can start laying eggs immediately and the foragers can quickly start bringing back nectar and pollen to store.

Lay some prepared wired frames (without foundation) horizontally on a sheet on the ground, then cut out a section of brood comb from the possum box colony and use string to attach the comb to one side of the wired frame. Select comb with brood and eggs over comb with honey and pollen. Continue this until you have removed the brood comb and then fill any remaining frames with any honeycomb, rubber banding in the same manner.

5.19: Instead of using string to attach comb to frames an alternative method is to use elastic bands — the thick elastic bands used by the postman are ideal for this.

5.20: If you plan to catch many swarms it will be useful if you make up some frames specially for holding cut out brood comb. Fence netting attached to frames that do not contain any frame wire make good comb holders. Attach the fence netting only to the top bar using cable ties or staples that can act as hinges.

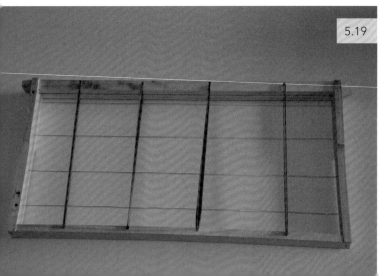

5.19

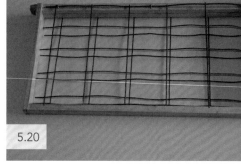

5.20

An alternative to using string to attach brood comb to frames, particularly if you are regularly removing established nests, is to make up some frames with wire mesh on the sides as shown in the photograph.

Attaching brood comb to a wired frame is an untidy business and it is unlikely that you will get the usual number of frames into the hive box or nucleus you are using, but this is not important. Any honeycomb left over can be placed in a sealed container and the honey extracted later as explained in method 1 in Chapter Nine on extracting honey. Do make an effort, though, to cut out all comb containing larvae and eggs for your prepared wired frames. If at all possible leave the hive with the frames of cut-out comb and attendant bees next to where the possum box was for a day or two. This will allow any residual bees from the possum box to find their new home.

An alternative method of transferring brood comb from a possum box to a hive is to simply cut out sheets of comb and to rest them vertically inside a hive. To stop the brood comb touching other brood comb place thin strips of honeycomb between them so that the nursery bees will be able to feed the larvae. Although this is a simple technique if the hive is not to be moved, moving a re-housed feral colony using this method will probably result in the comb falling over and many of the bees being killed.

If all has gone to plan, inside the hive the queen and the workers will be busily establishing a home. This, however, is not the end of the hard work since in a few weeks you will need to tidy up all the old brood frames, replacing them with new frames and foundation. What you do next will depend on whether you placed your cut-out frames in a nucleus box or in an ordinary eight- or ten-frame hive box. If you used an ordinary hive box prepare a second box, inserting into it frames with new foundation and some used drawn comb. Remove from its base the box containing the previously cut-out frames, placing this gently to one side on a level, flat surface. Take the newly prepared second box and place it on the now available base. Next, remove each frame in turn from the old brood box and shake all of the bees into the new brood box sitting on the base. When you do this it is essential that you take care to transfer the queen to the new brood box otherwise you will need to repeat this process in another week or two. Once all the bees have been removed, place a queen excluder over the new brood box containing the queen and bees and carefully place or slide the old brood box on top. Over the next three weeks all of the eggs and brood in the top box containing the previously cut-out comb will hatch and you will be able to remove these untidy frames and clean them ready for reuse. With the queen established in the new bottom brood box, the workers will start drawing brood comb and the queen will recommence laying eggs.

After a couple of weeks, if there is no sign of brood in the bottom box, but there are eggs and larvae in the top box this will indicate that the queen remains in the top box and you will need to repeat the process described above until the queen has been successfully transferred to the lower box. Once the queen has settled into the new bottom brood box and all of the brood has hatched in the top box, it is time to remove the frames in the top box and replace them with new frames and foundation, magically changing it into a super and providing room for the new colony to expand. As well as the old frames, the top box itself may be messy with remnant comb with the result that the whole box may need a good clean before it can be used again.

If, after a period of time, the newly hived colony is overly aggressive, not storing any quantity of honey or the queen is underperforming — that is, not laying many eggs — you will need to consider re-queening. Obtain a replacement queen from a reputable queen rearer and the performance of the hive should improve markedly after about three to four weeks.

A newly caught swarm often needs some help to build up its strength quickly. The beekeeper can assist by feeding the colony with sugar syrup and/or pollen substitute for a few weeks until the bees can stand on their own six legs and look after and sustain themselves. See Chapter Seven on spring and summer management for instructions on feeding syrup to colonies.

When tidying up after catching a swarm or moving an established feral colony it is important to mask the smell of the pheromone that the bees have left behind. A lemon-smelling toilet freshener spray is often carried by swarm collectors for this purpose and this is applied liberally to the branches or other structure where the swarm had landed or the colony had made their home. If it is really difficult to get access to the area where the swarm or colony was, buy a bottle of almond oil essence and spray this where the bees were. Bees hate the smell and within a minute or two, depending on the accessibility of the colony, the bees will have disappeared looking for a less smelly place to live.

MOVING HIVES

It is quite safe to move a hive inside a car if the hive entrance is well sealed and the lid, boxes and base are held together securely. The best way to do this is to use an Emlock fastener or to tie the hive together with ratchet straps of the type used for holding loads securely on the trays of utility vehicles. Duct tape may also be used to further seal the joins where the brood box and lid or base meets. Many new beekeepers feel safer moving a hive on a trailer rather than inside their car. The hive needs to be securely tied or strapped to the front of the trailer so it will not shift or move around during the journey. The gentle rocking of a carefully driven vehicle during transport of the colony has a calming effect on the hived bees and they are generally quite placid during the move.

When I move a hive using my car or trailer I always carry a veil and gloves to deal with any emergency that may arise, such as bees escaping due to a box or lid becoming displaced.

Bees in an established hive should be prepared for moving after dusk when the foragers have returned home and all of the bees are inside the hive resting. Sealing the hive entrance can be done by inserting a damp sponge or rag, tacking a strip of wood over the entrance, using a perforated metal entrance closer or plenty of duct tape. Make sure though that there is not a long tail of sponge or rag hanging out of the entrance that could get caught and pulled out. Whatever method you choose, if you try to seal the hive during the day when many foragers are out, you will find that very soon there will be hundreds of returning bees attempting to get back inside the hive. This will make completing your job more difficult and rob the colony of bees who will be lost if the hive is moved without them. A collection of agitated bees will also make moving the hive more difficult.

5.21: When moving a hive, the entrance may be sealed with a piece of metal.

5.22: Make sure, though, that you seal the entrance after dark when all the bees are inside the hive or the front will soon be covered with returning bees, making it difficult to safely move.

When choosing the entrance closure method, you should think about how long the colony is to be closed in and consideration should also be given to the temperature on the day of the move. If the hive is travelling for a short local trip in coolish weather the sponge or wooden seal can be used. If the hive is to travel some distance a ventilated entrance closure should be selected to allow the bees to keep air circulating throughout the hive. For any trips on very warm days or for long distances the wisest course is to use a ventilated travelling lid, which is a strong wire screen supported by a wooden surround. If the screened lid is used in full daylight, place a hive mat over the screen to limit the amount of bright sunlight and heat entering the hive. The base of the hive can also be modified to allow a better air circulation.

5.23

5.24

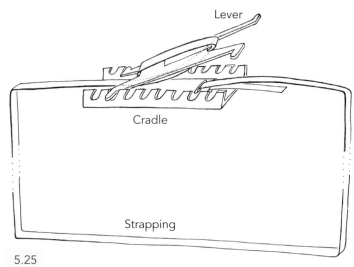

Lever

Cradle

Strapping

5.25

5.23: These hives have ventilated lids allowing the colony to breathe when they are moved long distances.

5.24: Emlocks are used to strap together a hive ready for moving. Note that soft foam has been pushed into the entrance to completely seal the hive. If the hive is only going to be closed for 20 minutes there should be sufficient airflow through the vents in the lid for the colony to breathe.

5.25: Emlocks are assembled as shown. First attach the strap to the cradle by bending it and pushing it through one of the slots. Next the other end of the strap is wound around the handle.

It is better to wait until after dark before sealing the hive even if you plan to move it later that evening or very early the next day. As long as the day is not hot, it is safe to keep bees overnight and during the next morning in a well-ventilated and sealed hive. They must not, however, be left in the baking sun for even a short period of time. Ensure that they have fresh air to breathe and water to drink. Bees are living creatures; once they are confined in a hive it is impossible for them to regulate the inside temperature and they can very quickly overheat and die. Not only do you risk losing brood and bees but there is the added risk that the frames will melt down all over the colony.

Once the hive has arrived at its new home it should be moved into its permanent location before the entrance is opened. Prior to doing this, the site should have been selected and checked for suitability, with any brick bases or stands previously prepared. Preferably after dark or very early in the morning when the bees are still resting, move the hive quickly to the new location, remove the straps holding the hive together, and then remove whatever is being used to seal the entrance. If moved after dark and the entrance sealer removed, a handful of bees will slowly make their way out to check what is going on before returning inside the hive. Early the next morning you will see many bees flying around the hive taking orientation flights before leaving to look for nectar and pollen.

Many beekeepers say that hives need to be moved further than 4 kilometres from their previous location otherwise the foragers will quickly return to their old home site. My experience is that I have moved hives very short distances, even 50 metres, without experiencing a great loss of bees. Some of the foragers may return to their old site although I would not let this deter you from moving a hive a short distance if it is necessary to do this, particularly if the circumstances surrounding the move are urgent.

Emlocks, while not an essential item of equipment, are very useful when you need to strap a box together in order to move it. Straps for Emlocks can be made from polyester webbing, galvanised steel, stainless steel or Colorbond coated steel.

Polyester webbing is the cheapest and also has the advantage that the flexibility of the strap makes it very easy to use. Perhaps the major disadvantage is that it cannot be left out for months in the sun and weather as the fabric will slowly disintegrate. For a hobby beekeeper who moves hives occasionally a polyester strapped Emlock is a good choice and when the strap deteriorates it is both easy and cheap to replace.

Galvanised steel is a more robust material than polyester and can be left out in the sun for years without disintegrating. For this reason it is a strapping of choice in situations where hives are to be left outside for long periods.

Galvanised and Colorbond steel are a lot stronger than polyester and the strap is unlikely to break during use. It does, however, have the disadvantage that it is not as flexible as polyester and is a little more difficult to use when strapping together a hive.

Professional beekeepers often use stainless steel strapping for their Emlocks, not because it is more rust proof than galvanised steel, but because it stretches and is less likely to break after constant use. Stainless steel strapping is slightly more expensive than galvanised steel, but is worth the extra cost due to its superior performance over a protracted period of time.

Although I have never had an Emlock come apart on me, for peace of mind I always use two Emlocks to hold together a hive that I am moving. This extra insurance makes sense when you have 50,000 or more agitated and angry bees half a metre away from you inside a car.

GENERAL MAINTENANCE AND INSPECTIONS DURING THE FIRST YEAR

Between spring and late summer, the peak period for honey flow, the beekeeper needs to regularly inspect the hive to check on the health of the colony, to manage swarming, and to see if honey needs to be extracted or a super added or removed. This is an ongoing procedure and is straightforward once it has been performed a few times.

Chapters Seven and Eight on spring and summer management and autumn and winter management provide detailed recommendations on how to approach the

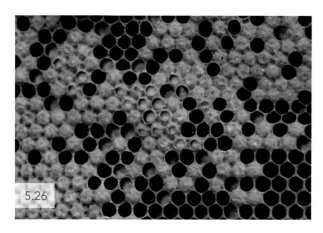

5.26

5.27

5.28

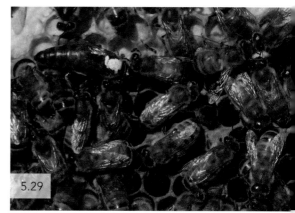

5.29

5.26: Larvae and capped brood.

5.27: A frame full of healthy capped brood.

5.28: A new bee emerges from her cell. This is seen frequently inside a hive and is one of the delights to watch as a beekeeper.

5.29: Queens are usually difficult to find. When you find a queen always try to mark her so that she will be easier to find during subsequent visits.

5.30: Queen cells in a hive are a sign that either the colony is about to swarm or they are replacing a poorly performing or missing queen.

5.31: There needs to be plenty of pollen inside a hive if the colony is to raise healthy brood. Since a single floral source seldom provides the range of proteins that a colony needs to rear healthy brood, look for cells that contain a range of pollen colours since this indicates that the foragers are collecting food from several types of flora.

5.32: The capped brood cells show an irregular or patchy distribution of brood; this indicates a diseased colony and you need to determine the cause of the disease and treat it accordingly.

5.33: Castes of bees. From left to right: female worker, male drone, female queen.

inspection of your hives. The instructions detailed in these chapters are suitable for beekeepers during their second and subsequent years of looking after bees. In the first year of your hobby, however, the new beekeeper will need to follow a slightly different course for inspections. This is because the beginner will not be familiar with typical colony behaviour or what to look for inside a hive. As a result, inspections during the first year should be aimed at gaining confidence and experience in opening the hive and learning about the colony of bees inside.

Confidence is important and you will need to inspect the colony often so that you gain this quickly and feel safe from stings wearing your protective clothing. You will need to open your hive and look inside probably once a week for the first few months. (Later, the frequency with which you open your hive for inspection can be reduced to about once every three weeks during the spring and summer.) This weekly visit need not involve removing any of the frames during every inspection although gaining familiarity with this task will help in quickly developing your skills and confidence as a beekeeper.

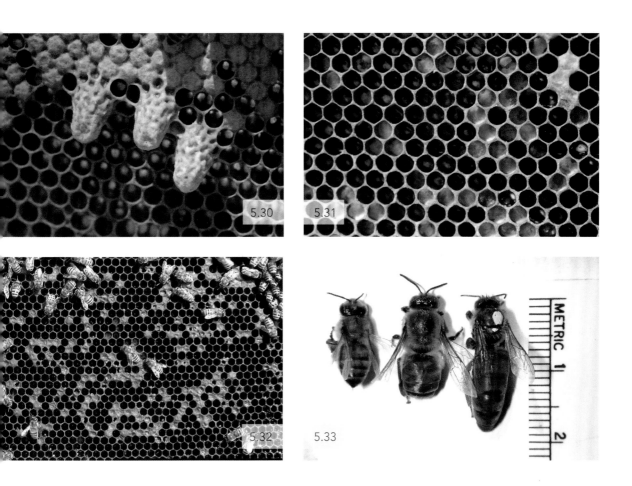

5.30

5.31

5.32

5.33

Another aspect of gaining experience as a beekeeper is honing your powers of observation of the movements of the colony outside the hive. Watch the foragers come and go to see if they are bringing in pollen on their legs. Can you see worker bees feeding each other with nectar at the hive entrance? This is a process called trophallaxis. Observe the general look of the hive when things are going well, so that when they are not you will be able to see the difference even before you open the lid.

Similarly, the new beekeeper has to become familiar with the look and feel of the bees inside the hive. If anything changes later during the year it is then much easier to identify what is different and to seek advice.

If you bought the hive as a viable colony do you know the makeup of the hive you are inspecting? Is the colony in a brood box with an excluder between that and a honey super? Alternatively does the hive have two boxes but no excluder? We will assume for the purposes of explanation that you are inspecting a brood box, although in practice it will often be a honey super that will be the first box facing you when you remove the hive lid.

After successfully lighting your smoker, before you approach the hive, make sure that the fuel is well alight and that the smoker is blowing a thick, cool white smoke. I can't stress this enough. Many new beekeepers do not ensure the smoker is well alight and smouldering plenty of cool smoke.

Before any attempt is made to open the hive, first puff smoke into the hive entrance. Then move to the rear or side of the hive and puff another couple of times under the lid as you gently slide it forward enough to give the smoker snout access. Wait about 15 seconds for the smoke to take effect, take off the lid and puff more smoke under the hive mat. This calms the bees and drives them down into the hive box. Remove the hive mat and quickly conduct the first visual check by looking down between the frames into the hive.

It is only necessary to use two or three puffs of smoke to subdue bees. Beginners often use excessive amounts of smoke, which the colony can take many days to recover from. If you are dealing with very aggressive bees, it may be necessary to wait for 2 or even 3 minutes after smoking before opening the hive.

Then, starting with the second frame from one side, remove the frame and brush or shake most of the clinging bees back into the hive. Visually check both sides of the frame for eggs, larvae, brood patterns, capped or uncapped honey and pollen and make a mental note as your inspection proceeds if there is anything that concerns you and which you may need to discuss with an experienced beekeeper or do some research on. A nucleus box is a useful aid since you can temporarily store frames in it while inspecting your hive.

5.34

5.35

5.36

5.37

5.38

5.34: Inspecting a hive. First remove the second frame in the hive as shown. The first frame next to the hive wall may be attached to the wood and become damaged if it is removed first.

5.35: Next prise the first frame, next to the hive wall, away from the wood using the flat end of the hive tool. Once the frame has been detached from the hive wall it can be removed and inspected.

5.36: American hive tools are designed differently to the 'J' style or Australian hive tool. The hook end is first used to separate the frame to be removed from its neighbouring frames and then it can easily be removed.

5.37: When I remove frames from a hive I usually store the first frame on the ground resting against the brood box. With one of the frames from the hive removed, there is now plenty of room inside the hive to move, remove and inspect the remaining frames.

5.38: An alternative method of storing frames during a hive inspection is to rest them on a frame holder attached to the super.

I have assumed at this early stage that you are not familiar with disease in the hive so have not suggested a check for this. If, however, you have done your homework and feel you know what to look for this should also be part of your inspection.

Once you have removed and inspected the first frame you can either place it on the ground leaning against the hive, use a frame holder attached to the hive to temporally hold inspected frames, or place the frame in an empty super or nucleus box beside the hive. When you start it is a good practice to adopt the spare box method of storing your frames during hive inspections. It will keep your frames clean and safe. Frames leaning against the hive sometimes fall over or you can accidently put your foot through them when you are engrossed in your hive inspection.

Next, use the flat end of the hive tool to prise the first frame (the second you are removing from the hive) away from the hive wall and remove it for inspection. Again, to give yourself plenty of room inside the super or brood box you are inspecting, make sure that after you remove this frame it is placed outside the hive leaning against the hive wall, hanging on the frame holder attached to the hive or in the spare super beside the other frame. The reason for starting your inspection with the second frame rather than the first frame next to the hive wall is because the first frame will often be stuck with honey or wax against the hive wall and pulling it out with another frame flush beside it may be difficult, resulting in a damaged frame, broken honeycomb and honey leaking out inside the hive.

Next, gently prise the third frame away from the fourth frame and remove it for inspection. After inspecting the third frame, return it to the hive slightly away from where it was taken, allowing you plenty of room to inspect the fourth frame and so on. Once you have inspected all of the frames, return them to their original positions. As you gain in experience it will seldom be necessary to inspect every frame in a super or brood box since a lot of information can be obtained by looking down between the frames and by removing for inspection two or three frames randomly from either box.

If you are inspecting frames from the super all you are looking for is a full frame with about two-thirds of the cells capped with a wax layer, which will mean that the honey underneath the cappings is ready for extraction. If most or all frames in the super are almost fully capped, the frames are ready for extraction. If only one or two frames are full, you can remove these frames and extract them by hand or, if you own an extractor or want to wait until you have a greater number of full frames, you may choose to store them. If you choose to do this you will need to freeze frames to avoid any chance of Wax Moth larvae hatching and destroying the harvest. See Chapter Nine on extracting for details of how to do this. You will need to replace any full frames you remove with empty frames containing new foundation or empty comb. Alternatively, you can leave the capped frames inside the super for a week or two longer until the remaining frames are also capped and all the frames in the super can be removed and extracted at once.

The only times I fully inspect every frame in a brood box are when I am looking for the queen or I suspect that the colony is diseased and I need to determine if this is indeed the case and the extent of the disease. Also look for the queen when you remove frames from the brood box; she can be difficult to find, particularly for a beginner who has little experience spotting queens. When you find her it is well worth the effort of marking her thorax so that she will be easier to find when you need to. Bee supply stores sell a small round cage to hold the queen during this process and a marking pen to add a mark to her thorax. A hobby beekeeper with a few hives can use any colour pen but a larger apiarist may wish to go with the internationally recognised queen colours that signify at a glance the year the queen was introduced to the hive.

While the hive is open, take the opportunity to clean the hive walls and frames of any burr comb and also remove any burr comb from inside the lid of the hive. Burr comb in the lid can be minimised by placing a hive mat above the top set of frames under the lid, making sure that there is sufficient gap around the mat for air to circulate. While the hive is open, also check for damaged frames that may need to be repaired or replaced.

For a detailed list of activities to be undertaken during each inspection in subsequent years see Chapters Seven and Eight on spring and summer management and autumn and winter management.

SUMMARY

- Bees can be purchased as part of an existing hive, caught when a colony has swarmed, or purchased from a beekeeper as a nucleus hive. Each of these methods is popular so you will need to choose which method of obtaining your first bees suits you best.
- Once your colony is housed permanently on its site regularly inspect the hive both to gain confidence that your protective clothing will protect you from bee stings, and also to learn the ways and temperament of your colony. An important part of the beekeeper's work is to regularly inspect their hives to check that all is well with their bees and to remove any full frames of honey that need extracting. To become proficient and confident at these activities takes practice, although interest and enjoyment are also reasons for performing them regularly.

6.

6.1: Try to position your hive so that the entrance faces north-east into the morning sun. This is important as an entrance that faces the morning sun is more likely to allow the hive to remain dry and will thus help minimise disease.

6.2: When deciding on the optimum location for your hives there are several things that need to be taken into consideration: morning sun for the hive entrance, shade during the midday heat, a water source nearby and easy access for the beekeeper.

6.1

Locating your bees

LOCATING HIVES ON YOUR PROPERTY

For a hive to remain healthy and productive, its location and orientation is important. Place the hive so that the entrance faces a north-easterly direction into the rising morning sun as this helps to keep the hive dry. It is important that your bees leave the hive as early as possible to collect nectar and pollen, particularly if there is competition from other colonies of bees living in the area surrounding your hive. You want your bees to get the lion's share of any available nectar and pollen and an early start allows for maximum foraging opportunity. Just like other insects, bees are poikilothermic (this is often referred to as 'cold blooded') and the warming rays of the first sun in cooler weather will increase the ambient air temperature around the hive entrance, encouraging the bees to begin foraging activity.

As previously mentioned, prepare the site prior to moving the hive. If you propose to make a brick platform or place the hive on a stand, have this ready in position prior to moving the hive on to the site. If the hive is to be placed on a rural property, ensure that it is raised off the ground so that there is no risk of long grass and weeds blocking the entrance. This can happen very quickly in times of peak growth like early spring. Bees do not like wind and areas that tunnel wind or are very exposed should be avoided for hive placement.

Whenever I move a hive to a new location I ensure that the hive floor is slanted slightly forward, allowing any water that has collected inside the hive to run out. If the

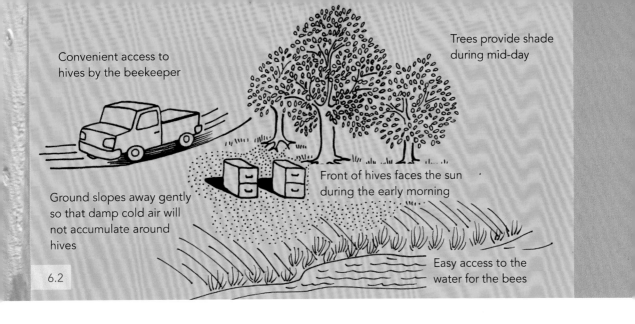

Convenient access to hives by the beekeeper

Trees provide shade during mid-day

Front of hives faces the sun during the early morning

Ground slopes away gently so that damp cold air will not accumulate around hives

Easy access to the water for the bees

6.2

hive were to slope slightly backwards water would pool at the rear of the base, causing the bees to endure wet damp conditions or to drown.

While the hive entrance should face the morning sun, it should be covered by shade during the height of the mid-summer day, otherwise the hive will become too hot and the bees stressed. During the winter the opposite occurs and the hive should not be covered by shadow during the day since the bees will consume too much honey keeping the colony warm. Damp, shaded positions encourage mould and condensation in hives, neither of which should be encouraged for the health of your bees. Satisfying all requirements is not easy and the conditions you and your bees face will largely depend on which part of Australia you live in. Some hobby beekeepers find they need to keep hives in a different location in the summer than in the winter in order to keep the hive in optimum conditions for the bees.

Often, the biggest concern for the new beekeeper is how near to the house a hive of bees can be kept: will children be in danger, will pets be stung and will the bees bother neighbours? Many beekeepers, particularly in inner suburban situations, keep their bees only a few metres from house entrances, paths or from recreational areas on their property. In practice, I would not recommend this, although

6.3

6.3: When you position your hive, ensure that it is sloping slightly towards the front to allow condensation to run out.

much depends on the temperament of the bees and how confident you and your family are around them. My hives were kept for many years only 10 metres from my back door without any incidents. If the bees in any of my hives became particularly aggressive I simply re-queened with a gentler queen and the problem was solved in less than six weeks. In my own situation my family were not frightened by the presence of bees and there were other areas on the property where my children could play away from the hives. To obtain a queen with a better temperament contact a queen rearer and ask how passive their queens are. Alternatively members of your local bee club or a reputable bee supply store will be able to recommend a queen rearer, usually from personal experience or from customer feedback.

Generally I have not found bees to be a problem with smaller animals and, over the years, I have spoken with many beekeepers who have also kept a variety of other domestic animals with no harmful incidents. To sound a precautionary note, when you introduce bees on to your property be aware that young puppies and kittens can suffer badly from bee stings, particularly if their breathing passages begin to swell, and a visit to the vet for an antihistamine injection may become necessary. Cats generally avoid a hive but older dogs might need to be stung before they, too, learn to stay away. Some beekeepers keep their hives in the same run as their hens and bantams and the two usually live contentedly side by side.

Probably one of the few times that you need to make special arrangements for other animals is if you plan to keep hives in the same paddock as larger animals such as horses and cattle. Larger animals may see the hive as a rubbing post against which to ease an itch. In these instances, and in situations where there are other wild animals such as kangaroos or wombats with access to your property, it is better to fence off the hive with a small, inexpensive fence perhaps made out of star pickets and wire rather than risk the wellbeing of larger animals and your bees.

I think it is fair to state that one of the biggest concerns for the urban hobby beekeeper can be immediate neighbours. Amazing as it may seem to the dedicated beekeeper, some people just do not want bees near their property. The majority of people will suffer an uncomfortable topical reaction to a sting and if a particularly vascular part of the body like the face is affected swelling can be severe. Nevertheless this should not be confused, as it often is, with a systemic allergic or anaphylactic reaction to a sting. Severe allergic reaction to bee venom occurs in roughly 3 per cent of the population, however local council health inspectors often joke that judging by the number of complaints raised by residents in their community they must have the most allergic population in the country. Allergy levels seem to increase in an area when some residents realise that even a not-so-near neighbour is keeping bees! Generally, the average person views the bee as a nasty insect that stings; in fact, many are unable to tell the difference between a bee and the more aggressive European wasp. European wasp and bee stings cause roughly the same amount of pain for a person, however wasps can repeatedly sting and are usually more of a nuisance since they are strongly attracted to barbecues, soft drink cans and food, while bees are usually found only on flowers or in the wrong place looking for a water source.

As a responsible beekeeper you have a duty of care to your neighbours and should do all that you can to ensure that your bees do not become a nuisance to others. This will involve you re-queening, if necessary for a calm strain of bee, and showing some consideration in the timing of your hive inspections; that is, if you live in close proximity to your neighbour don't open your hive when they are gathering for a barbecue or other outside function. Never forget the peace offering or the shameless bribe of the results of your bees' golden harvest.

To ensure that beekeeping does not overly inconvenience neighbours, each state and territory has developed its own Apiary Code of Practice. You can Google your relevant code of practice to access the information or call your local council. In Victoria, where I live, although the Apiary Code of Practice is a Victorian Government law, it is administered by local councils through their planning codes. Each code of practice will detail how best to keep bees so that they do not become a nuisance to neighbours or the public. If there is a complaint about your bees and you can show that you are following your state's code of practice, most apiary inspectors will support you in any dispute. With few exceptions most urban dwelling beekeepers can keep at least two hives on their property. To confirm the number of hives permitted in your own area you will need to check your state or territory's Apiary Code of Practice or contact your local council.

Although regulations differ slightly around Australia most specify how close a hive can be placed to a neighbouring fence. This is usually around 3 metres from the adjoining fence line. Alternatively, there needs to be a solid, impenetrable barrier 2 metres or higher at the dividing line between the two properties. This barrier can be a solid fence or bushes, and dense trees also seem to suffice. Again the requirement in your own area will be specified in the applicable code of practice.

It is my experience, and I believe it is the experience of most beekeepers across Australia, that state government apiary inspectors are helpful and will work positively with beekeepers to minimise the impact of beekeeping on neighbours. Although this is also largely the case with local council inspectors, some councils or local authorities have passed laws restricting the keeping of bees within their jurisdiction. This can be a problem for hobby beekeepers, since the majority of bees in an area will come from other colonies, but if there are any incidents near to a beekeeper's property their bees will be blamed and rational discussion may not change a local council's view. Some urban beekeepers have removed their hives altogether as a result of council actions rather than face the costs of a court case. It is always worth checking your local council's beekeeping policies rather than assuming that because your state allows you to keep bees on your property, your local council will also allow this.

HIVE POSITION IN SUMMER AND WINTER

Even though the summer heat is unlikely to melt honeycomb, it can cause it to soften and sag inside the hive, particularly in areas of Australia that suffer from extreme summer temperatures. The build-up of heat inside the hive can be extreme, causing distress to the bees and resulting in poor conditions for them to raise brood. Winter cold I find easier to manage by insulating the hive with polystyrene sheets on the lid and by providing plenty of reserves of honey and pollen for bees to access when they are unable to leave the hive in search of food. I have also of late experimented with a wooden lid that telescopes over the sides of the existing hive lid but is supported above it, creating a large airspace between the two lids. This creates effective insulation in both summer and winter conditions while not impeding any airflow around the hive.

Feral bees prefer to make nests high above the ground. To help make my hives feel more natural to the colony, if I can, I raise them off the ground about 30 centimetres. This has the advantage that a raised hive will:

- allow better air circulation around and through the hive resulting in less condensation
- improve air circulation which leads to healthier bees and a dryer hive; generally *Nosema* or Chalkbrood will be reduced if circulation inside the hive is improved
- slow grass growing around the hive and over its entrance during your absence, which can restrict or even prevent access for the bees.

I try to raise the base of my hives roughly 30 to 40 centimetres off the ground; any more than this will make inspection and removing supers difficult.

URBAN BEEKEEPING

There is a growing interest in beekeeping by people living or working in cities and this has resulted in many people, who live in very small townhouses or even apartments with balconies, realising that keeping bees is a hobby that they can participate in and enjoy. Even restaurateurs and other business people with access to roof space in inner cities are finding that keeping bees is a worthwhile activity. Many people believe that bees and cities do not mix, that there are insufficient flowers for food, or that pollution is too high to be good either for the bees or the resulting honey. This view is far from the truth and inner-city beekeeping often results in a lot more honey and healthier bees than rural beekeeping. Why is this?

Although most people think that bees and rural areas go together, in fact, much of rural Australia is dry and supports bees only during certain times of the year. This is the reason that many professional beekeepers here transport their hives many hundreds of kilometres every few weeks to follow the flowering patterns of the gums or other nectar producing plants. If professional beekeepers kept their hives in a single rural location the amount of honey they extracted each year would more than halve compared to moving them around. Indeed, Australian professional beekeepers together with their North American counterparts are almost unique in practising migratory beekeeping by transporting their hives

6.4: I raise my hives off the ground about 30cm wherever possible. This stops the entrance from getting blocked by long grass as well as allowing air to circulate under the hive keeping it cooler. Oily rags are tied around the legs to stop ants entering the hive.

to follow flowering patterns. In other countries hives may be moved to take advantage of one floral source or another but not to the extent or for the distances they are transported both here and in the United States. For example, beekeepers from North Yorkshire and from parts of Scotland move their hives to take advantage of the seasonal flowering of heather.

During dry months and particularly in times of protracted drought, rural areas may suffer water shortages with the result that many migratory beekeepers transport water considerable distances to their bees during these times.

Bees kept in urban areas are not faced with the same difficulties. Even in the largest Australian cities there is an abundance of well-kept parklands, botanic gardens, rivers and many homes with well watered gardens of exotic flowers and plants, many of which honey bees thrived upon for thousands of years, long before they were introduced into Australia.

Rural areas often contain large agricultural areas and a lot of these are based on monocultures, or growing only one type of plant in an area. In monocultural areas, bees may suffer compromised nutrition and a classic example of this in Australia is the vast almond orchards of north-eastern Victoria. Monocultural planting is not good for bees since it does not provide a sufficiently balanced or varied diet. Bees need a variety of nectar and pollen sources for optimum health. In contrast, urban areas offer

6.5/6.6/6.7: Hives on rooftops.

6.8: Hives on top of Federation Square in Melbourne.

6.9: A hive in Hamilton Gardens, New Zealand.

6.10: A beautiful location for a hive in Tasmania.

a plethora of plant types for bees to feed from, both native bees and honey bees, so their nutrition is far better. Even pollution is not a problem as most urban hives are either kept on rooftops in the central business district or in quieter streets, well away from car and truck fumes. In fact, pesticide use in rural areas is generally far more of a concern than pollution in urban areas.

Many hives are kept on rooftops in both large and smaller cities, particularly by businesses such as restaurants and hotels. The main consideration is safety and ease of access for the beekeeper to both place the hives on the roof and to inspect and remove frames for extraction. Wind may be strong on exposed rooftops and a sheltered location for hives will need to be chosen.

Interestingly, not only do urban bees collect more honey than their rural cousins, urban bees usually start to build up numbers in the spring sooner than country bees, giving them a head start when collecting nectar for honey production.

6.11: Hives kept in the country.

FEEDING BEES WATER

There is a legal requirement that beekeepers must make water available to their bees. If there are no rivers or dams nearby, say within half a kilometre, an artificial source needs to be provided and one of the many ways to do this is to use a large chicken feeder. When I use a chicken feeder or other container to provide water for bees I make sure that there are sufficient stones, twigs, or floating objects in the water for the bees to stand on and access the water without drowning. I have heard tales of buckets of water being placed by beehives and quickly filling with drowned bees. It is not obvious to some people that bees cannot walk on water and must be provided with landing platforms. Bees are not fussy about the quality of water and seem to prefer either dirty

6.12

water, say from a drainpipe, or swimming pool water, perhaps because it contains more minerals or salt. The problem with swimming pool water is that it is often your neighbour's pool that the bees like visiting and this can lead to strained neighbourly relations. To minimise this, ensure that there is plenty of water near hives for bees to drink, as once your bees fix on your own or a neighbour's pool as a source of water it is often difficult to persuade them away from it.

The appetite and need a colony of bees has for water during a hot summer day can be astonishing. On hot days during a prolonged drought I regularly feed my bees 5 litres of water using a top feeder every two days. I find that mixing a little sugar with the water makes it much more palatable to the bees.

SUMMARY
- Decide beforehand where you are going to locate your hive and prepare the site in advance. As well as selecting the best location, this may include cutting the grass, generally tidying the site, and building a stand that the hive can be placed on.
- In the interests of neighbourly relations take care when positioning your hive that it is not too close to an adjoining fence. The regulations of most states and territories will require a distance

6.13

6.12: Using a chicken feeder to provide water to bees.

6.13: Bees drinking water off the ground.

of around 3 metres in the absence of either a solid barrier or one provided by trees and shrubs.

- The flight path of your bees should be considered and the hive positioned to avoid bee collisions with neighbours and people on public sidewalks. On your own property your hive should be placed away from children's recreational areas and commonly used access paths to clotheslines and outbuildings. If you follow your state's Apiary Code of Practice and there is a dispute with another person, the local apiary inspectors or council officials will usually be on your side in any dispute.

- Urban beekeeping is gaining in popularity and, as surprising as it may seem, bees kept in urban or central business district areas fare better than rural or country bees. There is a legal requirement that water needs to be provided for your bees. If there is a natural pool or stream nearby this should be sufficient, otherwise you will have to provide a reliable water source. Bees can safely be kept around most animals and a gentle colony can be kept on a residential block with few concerns provided some care is taken with hive placement. It also helps if your immediate family accept and enjoy your hobby.

7.1

Spring and summer management

Even though the management of your hive will be discussed on a season-by-season basis in the following chapters, the situation in your particular hive will strongly be influenced by what happened in the previous season, or even earlier. The health of a colony during spring is dependent on its management during the previous autumn and winter. Also, any adverse effects on a colony during a season will have ramifications well into the future if not promptly managed by the beekeeper.

As a result, when reading these two chapters do not take each season in isolation, but take into consideration the health and productivity of the colony during the previous season as well as where you would like them to be in the following three to six months.

The climate in the various states and territories of Australia differs enormously and within these variations there are further localised climates that will affect the beekeeper's management of hives throughout the various seasons. This will mean that there are differing activities for beekeepers in various regions at any one time during the year and these will depend on the seasonal conditions of the particular area. Chapters Seven and Eight on spring and summer management and autumn and winter management will correspond to these seasons in each region of Australia, although the actual timing of these activities among beekeepers in various areas will vary.

Preparing for the massive expansion in the number of bees during the early spring and then, six months later, preparing for the winter shutdown during the late autumn

7.1: Although almonds provide bees with a plentiful supply of pollen, they are a poor source of nectar and bees are unable to live off almond flowers alone.

are two of the most important management activities that a beekeeper faces each year. Fortunately, both are straightforward and, once performed during the first year, should become second nature in the following years.

September, October, November and December are the busiest months of the beekeeper's year. To minimise the amount of work that needs to be performed during this period make sure that all maintenance, building new hives and tidying of the apiary is performed during winter. Ensure that syrup feeders and any pollen/pollen substitute stores are at hand, ready for use. If you plan to re-queen a number of hives, pre-order your queens well in advance as they are often in short supply during the spring. If your hives need to be moved, say to satisfy pollination contracts early in the spring, make sure that you have the logistics planned well in advance. Any delay in the move or the build-up of bees in the hive may make the contract invalid or mean that you miss a honey flow in a region to which you were planning to move your hives.

To understand hive management you need to understand both the lifecycle of the bee and of the colony. You will also need to understand the seasonal pattern of the flora of your area. While there is little that you can do about the floral sources growing in your area except to move your hives to a more suitable region during lean times, you should gain sufficient knowledge to see in advance that trees are not budding as they should and be forewarned of problems.

There is a lot that you can do to modify the natural yearly cycle of your bees and this cycle will need to be managed if you are to get the maximum honey harvest from your colonies.

In order to maximise honey production you will need to:

- learn to recognise trees and plants and when they are in flower
- find out when the main honey flow occurs in your area and which plants and trees provide this
- manage your colonies so they build up a high number of foragers before the honey flow starts — the importance of high bee numbers prior to a honey flow cannot be overemphasised
- manage your hives after the honey flow so that they overwinter well and can build up quickly again at the start of the following spring.

If you want to maximise the amount of honey that your bees produce do not allow bees to build up numbers on the honey flow. Rather, build up the colony beforehand so the hive has the maximum number of foragers going into a nectar flow and is able to take the best possible advantage of the good conditions. To do this you will need to know from observation during previous years when the honey flow starts in your area and to feed syrup to the bees about four weeks before the honey flow commences. For most of the southerly regions of Australia this is usually during mid-August, but will differ in other parts of the country. In areas with little variance in summer and winter temperatures, sufficient nectar sources are often available during the winter to keep colony numbers high.

Correct timing of the colony build-up ready for the honey flow is critical. It takes three weeks for an egg to emerge as an adult worker bee and then a further three weeks for the bee to become a forager. The six-week period from egg to forager needs careful planning if the number of foragers is to be at maximum strength at the start of the honey flow when the highest number of bees needs to be out collecting nectar and pollen.

Most professional apiarists build up their colonies before spring by overwintering them in areas where there are winter-flowering plants rather than feeding with syrup. As an example, many beekeepers overwinter their colonies in the Mallee in north-east Victoria where there is sufficient nectar available to keep the colonies strong over winter.

Make sure all of your hives are strong. A large number of bees divided between a few weak hives will produce a lot less honey than the same number of bees concentrated in a single hive. Two hives each containing 30,000 bees will produce a lot less honey than a single hive containing 60,000 bees. Larger hives mean that proportionally fewer bees are required for in-hive activities, allowing more bees to become foragers.

INSPECTIONS

INSPECTING YOUR HIVES DURING EARLY SPRING

The last winter inspection of your hive, or even just watching the coming and going of the bees to and from your hive, should prepare you for what to expect during your first spring inspection. Examine your hives in spring when the colony has stopped clustering and bees are out flying. Do not open your hives if the air temperature is below 18°C as this may kill many of the young brood and weaken the colony.

Choose a warm day in late August or early September in which to open the hive and perform a thorough inspection of every box and frame, vital after the long winter months when you have been unable to do this.

During this inspection you need to consider the following:

- How strong is the colony?
- Are eggs or brood present? If they are, you know that the queen is there even if you do not see her.
- Is the brood pattern patchy or solid?
- How much food is left in case the weather deteriorates and the bees are unable to forage?
- Inspect for diseases such as American Foul Brood, European Foul Brood and Chalkbrood.

PROCEDURE FOR THE FIRST INSPECTION OF SPRING

Remove the lid from the hive and place it upside down on the ground next to the hive. Place the brood box on top of the upturned lid; if two brood boxes were used over winter place both boxes on the upturned lid.

Now that you have complete access to the bottom board or hive base do the following:

- Clean any debris off the bottom board and wash it down with a wet cloth.
- If there is an entrance reducer, it doesn't need to be removed during the first inspection but can be removed a few weeks later.

Inspect and clean each brood box if two brood boxes were used over winter. Now is the time to reverse them by placing the previous top brood box onto the base. This is because over winter the colony may have moved to the upper brood box and will not be utilising the full volume of the two brood boxes. By reversing the position of the two brood boxes you will break up the winter cluster, giving the colony plenty of room for the queen to lay her eggs.

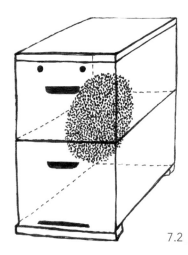

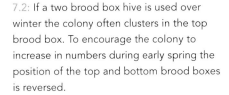

7.2

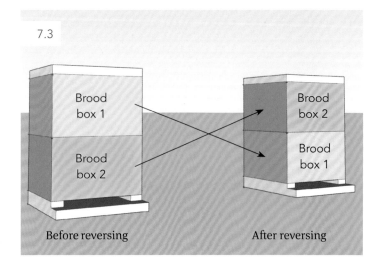

7.3

Before reversing   After reversing

7.2: If a two brood box hive is used over winter the colony often clusters in the top brood box. To encourage the colony to increase in numbers during early spring the position of the top and bottom brood boxes is reversed.

7.3: If you have kept two brood boxes over winter, now is the time to reverse them.

When inspecting the brood box be methodical and follow the steps outlined below.

- With the first brood box on the base, remove each frame in turn and check it for brood, honey and pollen stores.
- If brood is present, check that the brood pattern, both capped and uncapped, is solid. If the brood pattern is patchy this may mean disease, an infertile queen, or lack of food stores, and immediate action is required.
- Check to see that the queen is present. If you are unable to locate her look for newly laid eggs or very young larvae to confirm her presence.
- If you find a queen mark her as described in the chapter on rearing queens.
- If you decide the colony needs re-queening it is usually too early to buy a new queen in September and this activity will probably need to be performed later in the season, say in October or November.
- If you replace frames in a hive during early spring, preferably use frames with drawn comb kept over from the previous year so that the queen can start laying eggs in them immediately without the delay of the workers first needing to draw out the comb.
- If there are empty frames between the brood and honey frames, place the frames containing honey next to the brood but be careful

not to divide the brood by placing the frames of honey between the brood frames, effectively causing the brood to be divided into two parts.

- If the weather has been poor check the hives weekly for food. If insufficient food is present start feeding immediately.

Adding a new super

- Now that you have inspected the brood area, if the colony is sufficiently strong you can add a super on top of the brood boxes and at this stage, if used, a queen excluder between them. A sign that a super needs to be added is either that the brood box is full of honey and brood or that when you take off the lid bees overflow the sides of the brood box. The super that is being added should also preferably only contain drawn comb since we are coming into the main honey flow period and the bees should not spend time drawing out new comb but should instead be storing nectar.
- Often bees will not move up into a new super to store nectar. To overcome this, place some frames of brood in the super and the nursery bees will move up to look after the brood, encouraging other workers to follow.

7.4

7.4: Bees overflowing from the top of a hive when it has been opened is a sign of congestion and that the colony may be preparing to swarm. If this is the case add another brood box or super to provide the colony with more space to grow.

7.5

7.6

7.5: Burr comb in a lid is untidy and makes the lid more difficult to remove. A hive mat will minimise bees building burr comb in the lid.

7.6: Popsicle sticks under the lid provides additional ventilation for damp air to escape during early spring.

7.7: Keep the area around your hives tidy and equipment in good condition.

Clean the hive mat and lid

- Clean the hive mat.
- Remove and discard any burr comb from the inside of the lid.
- If the inside of the lid looks damp additional ventilation is needed. When you place the lid on the brood box or super, raise it up slightly by placing some matchsticks, small twigs, stones or Popsicle sticks under the corners. This will allow damp air to escape from the hive, keeping the inside drier.

Clean the area around the hives

- Control weeds and vegetation around the hive and next to the hive entrance. If you use a whipper snipper to cut the grass make sure that you wear protective clothing.
- Bees require a lot of water at this time of year so make sure that it is available.
- Remove any rubbish from around the hive and keep the apiary area clean and tidy.

An important aspect of hive inspection is swarm management. Many books and beekeepers say that to perform swarm prevention correctly the brood box needs to be inspected every week during the spring for queen cells. I find that a weekly inspection

7.7

disturbs the colony too much and seriously interferes with their normal activities of nectar and pollen collection. The bees can become sufficiently agitated that you may not be able to go near the hive for several days without getting stung. Later in this chapter I will describe other methods of swarm management and if these are followed a detailed inspection of your hives need only be performed every three weeks.

INSPECTING HIVES DURING LATE SPRING AND SUMMER

The second and subsequent stage of hive inspection of the season is a much-reduced version of the first stage of hive inspection explained on the previous pages. About every three weeks or less open the hive and check for the following:

Inspect the supers for honey
- Remove the lid and hive mat and inspect some of the frames in the super to determine how much honey is present. If most of the frames are two-thirds or more full of capped honey they can be removed ready for extraction. If the amount of capped honey is much less than this, determine how strong the colony is and how strong the honey flow is, to give you an idea of when to inspect next for capped honey ready to extract. See Chapter Nine on extracting for more details on removing frames for extraction.
- Clean any burr comb from the box or frames.

Inspect and clean each brood box
- Every second or third inspection check in the brood boxes for a laying queen. If you are unable to locate her, look for newly laid eggs or very young larvae to confirm her presence.
- If brood is present, check that the brood pattern, both capped and uncapped, is solid. If the brood pattern is patchy this may mean disease, an infertile queen or lack of food stores, and immediate action is required.
- The above two checks should tell you if the colony needs re-queening.
- If you replace frames in a brood box, preferably use frames with drawn comb.

- If the weather has been bad, check the hives weekly for food. If insufficient food is present start feeding immediately.

Adding a new super

- If the colony is sufficiently strong and there is a good honey flow you can add another super on top of the existing super. The new super should contain some drawn comb so that the colony can start storing nectar in it immediately if necessary.
- Workers often do not like moving up into a new super. To help the bees get used to the new super, try placing some frames of honey in it or even some capped brood.

Clean the lid

- Remove and discard any burr comb from the inside of the lid cavity.
- If the inside of the lid looks damp at the start of spring additional hive ventilation is needed. When you replace the lid on the brood box or super, raise it up slightly by placing some matchsticks, small twigs, stones or Popsicle sticks under the corners. This will allow damp air to escape from the hive, keeping the inside drier.
- Condensation inside a hive arises from bee metabolism used to keep the brood warm. If the weather is cold, ventilation is counter-productive if it makes it harder for the bees to thermo-regulate the brood. If the weather is extremely hot add additional ventilation by raising the lid slightly using matchsticks, small twigs, stones or Popsicle sticks under the corners. The lid should only be slightly raised for the duration of the heat wave.

Clean the area around the hives

- Control weeds and vegetation around the hive and, in particular, next to the hive entrance.
- Bees require a lot of water during the summer; make sure that it is available.

FEDSS — FOOD, EGGS, DISEASE, SPACE, SWARMING

A useful mnemonic to remember when inspecting hives is FEDSS: Food, Eggs, Disease, Space, Swarming.

- Food: are there sufficient stores of honey and pollen present?
- Eggs: if you are unable to locate the queen, eggs or very young

larvae in cells mean that she was present a day or two before.

- Disease: check for the signs of disease.
- Space: is there sufficient space in the hive for the queen to lay eggs or for the workers to store honey?
- Swarming: during spring and early summer, if the colony is congested or queen cells are present, split the hive to make two colonies.

MANAGING THE BUILD-UP OF BEES IN YOUR HIVE TO MAXIMISE HONEY PRODUCTION

To get the best possible honey production during a honey flow the number of bees in the hive needs to be at its maximum just before the flow. If the build-up is too slow and the number of foragers is still increasing at the start of the flow, the colony may miss much of the honey flow resulting in a poor harvest. If the colony strength is at its maximum too early the colony may swarm before the honey flow starts. Even if it does not swarm and the colony's population peaks well before the honey flow it will require extensive feeding to keep bee numbers up. At this point you may want to split strong hives and reunite them when needed to reduce the chances of swarming.

In the early spring, between roughly mid to late August and late September, monitor the number of bees in your hive. You can do this during a detailed inspection but as these are only performed roughly every three weeks a quick way to monitor congestion in your hive is to remove the lid and hive mat, and see how many bees spill over the top of the super. Also, when you remove the hive mat look down between the frames to see how many bees are in the top box. If the hive looks as if it is getting full of bees, add a second brood box or super to allow the colony to expand. As you increase the number of boxes, you can place the excluder either above the first or second box.

The trend these days among hobby beekeepers is to use two brood boxes and one or more supers for honey, so the excluder, if used, will be above the second box. If you are expanding the number of boxes in your hive from one to three, do not put the two additional boxes on at the same time but add the last box when the first two boxes are full of bees. Bees do not cope well with too much space and extra space also leaves the colony vulnerable to attack by pests such as Small Hive Beetle and Wax Moth. This is because there are insufficient bees in the available space to control these pests.

When adding a third box to an existing two-box hive, if possible place the third box in the middle between the brood and existing super box, rather than on the top. Bees often do not like moving up into a new box, particularly if both the box and frames are new. Adding a new box sandwiched between the brood and an existing super will encourage them to move up to the familiar scent of the existing super and overcome any reluctance.

MANAGING THE HONEY FLOW

It is not difficult to identify a honey flow using the following as a guide.

- Bees are busy collecting nectar and are usually not aggressive — although on Messmate Stringybark the bees become aggressive defending and protecting their colonies!
- White wax is being used to build new honeycomb.
- A frame of honeycomb tipped on its side causes a shower of nectar to fall out.
- The entrance of the hive is active with many bees coming and going.

Unlike honey, nectar takes up a lot of room in the comb until most of the water has been evaporated. For this reason before the honey flow starts there needs to be sufficient space in the hive for the workers to store nectar. An empty super needs to be added before the flow if the colony is to collect and store as much honey as possible. It is best to use drawn comb in the new super as the bees will be able to store nectar in it immediately. If a new super with only foundation is added, the bees will expend time and energy drawing comb and not take full advantage of the honey flow and as a result not maximise the harvest for the colony and the beekeeper.

If foragers are bringing back a lot of nectar to the hive the colony may become honey-bound. This occurs when most of the available brood comb is filled with honey, leaving little if any space for the queen to lay eggs. As a result the queen may stop laying and the number of available foragers will decrease with the result that little new honey is being stored. A honey-bound hive needs to be managed quickly by removing frames full of honey and replacing them with empty frames of drawn comb.

7.8: A frame that is two-thirds full of capped honey is ready to be extracted.

7.9: If forager bees return to the hive with too much nectar the hive could become honey bound and the queen has little room to lay eggs. This results in a rapidly weakening colony.

During a strong honey flow a super can be filled in one to two weeks, so add additional empty supers early to ensure that there is sufficient room for honey to be stored.

In order to maximise the amount of honey that bees can store in a super, place only seven frames in an eight-frame super or nine frames in a ten-frame super. This allows the bees to store more honey per comb and is easier for the beekeeper to extract.

MERGING COLONIES

If you catch a small swarm or have a colony that has decreased in size significantly over winter and is not strong enough to provide much honey, the best option is to merge it with another colony in order to produce a single strong hive. This is a straightforward task using the newspaper method. If both colonies are in separate hives, decide which hive is the strongest and leave it on its base on the ground, remove the lid, place two sheets of newspaper over the hive and make some horizontal cuts in the newspaper using a sharp knife. The cuts should not allow the bees to immediately move through the slits since the two colonies will fight if they merge too quickly. Next, place the second hive, without the base, on top of the first hive, leaving only the lid on top of the second hive. The only barrier separating the two colonies is two sheets of newspaper. Over the next few days the bees in both the top and the bottom hives will slowly eat through the paper, getting used to each other's scent as they merge together. This means it is very unlikely that they will fight, but will calmly join to produce a much stronger and healthier colony.

When merging two colonies it is preferable to kill the weaker queen before starting, but if I cannot locate her I just join them together and let them fight it out on the assumption that the stronger queen will win.

A strong queen heads a colony that collected the most honey during the previous season. She also lays a solid brood pattern with plenty of brood being raised; that

7.10: Two colonies of bees may be merged using the newspaper method. The lid of the first colony is removed and replaced with newspaper that has slits cut into it.

7.11: Next, the brood box of the weaker colony is placed above the newspaper.

7.12: After a few days the bees will have eaten through the newspaper and will have merged without fighting.

is, free from disease. A weak queen heads a colony that collected less than average quantities of honey during the previous season, does not lay a solid pattern of brood and may have a hive that is diseased. You may also prefer to eliminate the queen that produces the most aggressive workers.

FEEDING SYRUP

As part of managing hives for the spring build-up, many beekeepers feed their colony sugar syrup and/or pollen substitute. There is a general consensus among beekeepers that bees should be fed a sugar—water mixture at a 1:1 ratio during the spring and a stronger 2:1 mixture during autumn. Studies have shown that feeding bees the less concentrated 1:1 mixture in the spring encourages them to build up brood numbers and to go foraging. The stronger 2:1 mixture is better for encouraging the bees to convert the syrup directly to honey for use as food over winter.

Feeding syrup to bees is straightforward and there are several methods that are widely used. These are:

- frame feeders
- front or Boardman feeders
- top feeders
- placing a large can of syrup above the top super
- a Ziplock plastic sandwich bag.

Frame feeders are useful for feeding syrup to weak colonies since the feeder can be placed immediately next to the brood frames where the in-hive bees can easily find the syrup and carry it back to feed the brood. Frame feeders are also good when raising queens since their small size enables them to fit easily into four- or five-frame nucleus boxes. For stronger colonies, some of the available smaller frame feeders are not very useful since they do not hold sufficient quantities of syrup, are often inconvenient to refill and require the colony to be disturbed more often than is good for them, particularly in cold or wet weather. To stop the bees drowning in the feeder I insert Gutter Guard as a ladder for them to walk up and down on. Pieces of cork, straw or small twigs can also be used for this as they will float on the surface of the syrup and provide landing pads for the bees as the level drops.

Front or Boardman feeders are 0.5-litre to 2-litre reservoirs placed at the entrance to the hive with a small trough leading from the reservoir into the hive. Front feeders are easy to use, but have the disadvantage that they encourage robbing since the syrup is located at the hive entrance.

Top feeders, sometimes referred to as Miller feeders after their inventor Dr C.C. Miller, are to my mind the most practical method to feed syrup to bees. The top feeder

7.13

7.14

7.15

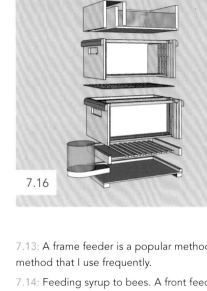

7.16

is essentially a very shallow wooden box; it has the same dimensions as a hive and sits under the hive lid. It consists of a large reservoir holding in excess of 5 litres of syrup. On one end of the feeder is a wide gap that allows the bees to crawl up from the hive and gain easy access to a small feeding reservoir. Top feeders have the advantage that they can hold a large amount of syrup and, if the feeder is designed carefully, the reservoir can be replenished without disturbing the bees or the beekeeper needing to wear protective clothing. To stop the bees drowning in the smaller feeding reservoir

7.13: A frame feeder is a popular method of feeding bees and a method that I use frequently.

7.14: Feeding syrup to bees. A front feeder is convenient to use but increases the risk of robbing.

7.15: Top feeders enable large quantities of syrup to be fed to a colony with minimal risk of robbing. I have had a 3.5 litre top feeder emptied every three days over about two weeks.

7.16: Schematic diagram of a top feeder and a front feeder.

I again use Gutter Guard as a ladder for them to walk up and down on. Pieces of cork, Popsicle sticks, straw or small twigs can also be used for floats instead of Gutter Guard.

Another popular method of feeding syrup is to put an empty super above the top super or brood box. A large tin or plastic container is placed upside down, resting on some thin strips of supporting wood. Large coffee cans are ideal for this purpose.

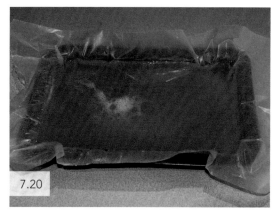

7.17: A can of syrup inside an empty super above the top super. Two sticks are used to raise the bottom of the can above the frames.

7.18: Ziplock bags of syrup with a slit across the top, resting on the top frames of the super, are an inexpensive way of feeding syrup to a colony.

7.19: Inserting a slit in a Ziplock bag to allow the bees to drink the syrup.

7.20: A block of candy in a baking tray can be used as an emergency food supply during the winter.

Punch some nail holes through the lid, about 1-millimetre in diameter, and the bees will drink the syrup through the holes. This is an inexpensive and convenient way to feed bees and is adopted by many larger beekeepers due to its ease of use. If the holes are kept about 1-millimetre in diameter a small quantity of syrup will flow out when the container is inverted but this will soon stop.

Ziplock sandwich bags are also an inexpensive and practical way to feed bees with syrup. Place the Ziplock bag on the hive mat or top bars of the uppermost frames and cut an opening in the top of the bag using a very sharp knife, razor blade or scissors.

There is only one rule for feeding bees syrup and that is to avoid any feeders placed outside the hive. The use of a central feeder placed away from the hive or away from a number of hives encourages robbing, feeds other beekeepers' hives or feral bees, is not hygienic and will lead to the spread of disease.

SUGAR CANDY

Syrup can be fed to bees during most of the year during the warmer months. During winter, however, particularly in colder regions like Tasmania and the higher mountainous regions of Victoria bees will not feed on syrup. During the winter months in an emergency I feed my bees with crystallised sugar candy, using the following formula.

Ingredients and equipment:
- 2.5 kilograms of white sugar
- 0.5 litres of water
- 4-litre aluminium/stainless steel saucepan
- Wooden spoon
- A bowl of cold water, big enough to sit the saucepan in or fill a sink with cold water
- A shallow baking tray or lamington tin lined with baking paper
- Plastic bags to store the slabs of sugar candy once it has crystallised in the baking tray.

Method:
Put the water into the saucepan on moderate heat and slowly add the sugar, stirring continuously. Bring it to the boil, then boil for 3 minutes. Remove the saucepan from the heat and stand in a bowl or a sink of cold water. While the mixture is cooling, keep stirring.

When the solution starts to thicken pour it into the baking tray or tin, scraping as much as possible of the solution out of the saucepan with the wooden spoon before the solution starts to set or candy.

When the slabs of sugar candy are cool they can be removed from the mould complete with any lining paper and stored in plastic bags until needed.

To use the candy I place it above the hive mat under the lid of the hive. The bees will find it here and eat it as they require. Candy is only an emergency feed if frames of capped honey are not available and cannot be used as a long-term feeding strategy.

FEEDING POLLEN OR POLLEN SUBSTITUTE

Syrup is fed to bees as a substitute for nectar so that they can build up stores of honey, albeit poor-quality honey. Honey is used to provide carbohydrates for the bees so that they have the energy to walk or fly, keep warm and to run their internal biological processes. This is only part of their needs, however, and for the colony to survive it also needs a plentiful supply of pollen. Pollen is used as a source of protein

7.21: Brushing pollen into empty brood cells to feed protein to the colony. This method of providing pollen to a colony is more efficient than sprinkling pollen over the top of brood frames.

and minerals, and is essential if brood are to grow and develop. Nursery bees eat pollen in the form of bee bread and convert it in the hypopharyngeal gland in their head into royal jelly. The royal jelly is then fed to larvae and, in this way, they obtain the protein they need to grow into healthy bees.

Pure pollen is more expensive than pollen substitute, which is typically made out of soy flour or yeast, although pollen is sometimes added to pollen substitute to increase the nutritional value of the food. Recently, irradiated pollen has become available from China and is growing in popularity among professional beekeepers.

Previously, the most common method of feeding pollen substitute to a colony was to mix the powder with sugar syrup to make a patty and to leave the patty above the hive mat for the bees to eat. This method, although convenient, has fallen out of favour since the patty attracts Small Hive Beetle and Wax Moth larvae that thrive in the protein-rich damp mixture. The preferred method today is to brush the dry pollen or pollen substitute powder directly into empty brood comb where it will last for a considerable amount of time. It is possible to place a teaspoonful of dry powder on the top of each brood frame and let some of the powder fall down over the bees. My experience with this method is that the powder, if not quickly consumed by the bees, will attract moisture and grow mildew. This will also happen to any powder that ends up on the base of the hive. As a result the base will need frequent cleaning so that it does not attract Small Hive Beetle and Wax Moth larvae.

SWARMS

MINIMISING SWARMING BY COLONIES

While swarming is a natural part of the reproductive lifecycle of the honey bee, there are steps that can be taken by the beekeeper to minimise it. These include:

- Ensuring that the hive is not congested with bees.
- Alleviating congestion by the addition of a further brood box or super. If two brood boxes are already in use on the hive, swap the position of the two brood boxes. Take the top brood box and place this under the existing bottom box, as this will provide more room for the queen to lay eggs. The top brood box is usually more

congested than the bottom box and swapping their positions will help to overcome this. If, however, both brood boxes are full and any queen cells are present you should artificially swarm or split the hive into two as described below.

- Making sure that there is plenty of room for the bees to store honey and pollen. During a strong honey flow the bees will store nectar in brood cells and can cause the hive to become honey bound. If there is no room for nectar to be stored, either remove capped honey frames or add another super.

- Regular re-queening with a young, healthy queen should minimise any swarming activity over the current season since queens seldom swarm until they are a year old. Another advantage of re-queening every year is that queens lay far more eggs during their first year, so the colony will be much stronger as a result.

- Selecting queens that are of a lineage less likely to swarm. Professional queen rearers raise queens that are selected for several attributes, one of which may be less of a tendency to swarm. Talk with several queen rearers about the traits that they select for and also with other beekeepers about their experience with commercially purchased queens.

- Using artificial swarming techniques such as the Demaree (see p. 126) method to control swarming.

- Providing sufficient ventilation to keep the hive cool during hot days and evenings.

If, despite your best efforts, you believe a hive is about to swarm you will probably be able to manage swarming using one of the techniques described below.

SPLITS

One popular method of managing swarming is to fool the old queen into thinking that she has already swarmed and is living safely in a new hive. This is called the split method and is described here. Next to the hive that is about to swarm place another hive, preferably one containing a few frames of old brood comb. From the hive that is about to swarm, remove about half of the frames that have brood in them, very carefully inspect each frame that you have removed to see if there are any queen cells on them. Be aware that a careful inspection is called for since some queen cells are not easy to see. If there are queen cells present, place all of the frames with queen cells into the new brood box that you have prepared, leaving the old queen in the old hive with the remaining frames of brood and honey. The old queen will think that she has swarmed and will remain contentedly in the old box, while the new box containing the new queen cells will develop into a new colony with healthy bees. Of course it goes without saying that you must replace the frames removed from the old hive with new ones, either with foundation sheets or, even better, ready drawn comb if you have them.

For more detailed information on splitting a hive see the split method of rearing queens in Chapter Eleven.

DEMAREE METHOD

Another popular and well-utilised method of managing swarming is the Demaree method, named after George Demaree who first published the details of the method in 1884.

When using the Demaree method, the beekeeper first separates the queen from most of the brood by placing the majority of brood into a second brood box that the queen is unable to enter. The two brood boxes are separated by a super and two excluders. If you were to look at the set-up you would see a brood box with an excluder on top. Then a super is placed on top of this with a further excluder where the lid would normally sit. A second brood box is then placed on top of the excluder and a lid placed on top in the usual way. The result is a central honey super with two brood boxes, one on the top and one on the bottom. This results in very little congestion and adds room for the queen to lay eggs. In essence, the colony believes that it has already swarmed.

Here are the basic steps:

1. Lift the brood box from the base.
2. Above the base place a new brood box filled with empty, drawn comb.
3. Remove the centre two frames of drawn comb and set aside.
4. Go back to the initial brood box and find the queen.
5. Place the queen and two frames of sealed brood in the centre of the new brood box sitting on the base, leaving most of the worker bees in the top box.
6. Place a queen excluder above this box.
7. Above the queen excluder, place one or more empty honey supers and then the original brood box where you found the queen. Push the brood nest together in the centre and put the two empty drawn frames from step 3 on either end of the new brood box.
8. Add your hive mat and lid.
9. After one week go through the top brood box and remove any swarm cells.
10. If necessary, the entire procedure may be repeated after nine or ten days if the hive continues to be congested.

7.22

7.23

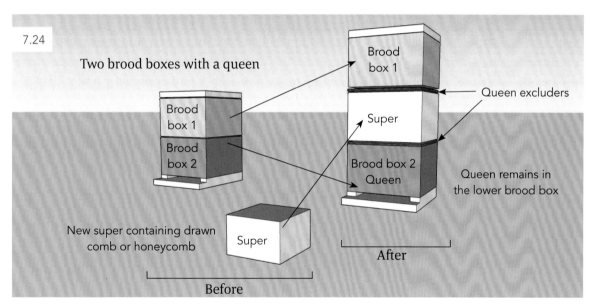

7.24

Two brood boxes with a queen

Brood box 1

Brood box 2

New super containing drawn comb or honeycomb

Super

Before

Brood box 1

Queen excluders

Super

Brood box 2 Queen

Queen remains in the lower brood box

After

Now that you have the Demaree hive set-up this is what happens.

- Most of the nursery bees stay with the brood in the top box and care for them.
- A small number of nursery bees stay with the queen in the bottom brood box and look after the two frames of brood that were placed in it.
- The older workers continue to forage for honey and pollen.

7.22: Demaree Method. This two-box hive is congested and is likely to swarm.

7.23/7.24: The queen and one or two frames of brood are left in the bottom brood box while most of the bees and brood are placed in the top box. A third honey super is placed between the bottom brood box and the top brood box, with two queen excluders placed at the bottom and the top of the honey super.

- The queen continues to lay eggs and has lots of room on new frames to do so.

This situation, while much like a hive that has already swarmed, has the major difference that both parts of the colony are in the same hive. You should therefore be aware of the following:

- As soon as the queen scent decreases in the top box, the bees may try to raise a queen from young larvae.
- You may need to destroy these cells or remove them to another nucleus hive.
- After the brood hatches, the brood cells will be backfilled with honey.
- In the end, the hive will not have swarmed, so it will contain lots of bees and lots of honey.
- The growing hive may once again develop the urge to swarm, which is why a second Demaree is often needed.

The Demaree method can be a very effective tool to use in swarm control, but as you can see, it is quite labour intensive. It involves a lot of manipulation, provides good opportunities to lose or damage your queen and involves a lot of heavy lifting. On the other hand, not only does the Demaree method prevent swarming, but you can obtain some queen cells in the process. For the Demaree method to be successful you also need to be able to find the queen. Every time you inspect the brood box of a hive look for the queen and if you find her make sure that you mark her so she can be more easily found next time.

One more important point: when you set up the Demaree hive be sure to remove any swarm cells that are already present. Any cell not removed may hatch and cause a problem within the hive.

PROVIDING HIVES FOR POLLINATION

Many professional and hobby beekeepers provide hives to horticulturalists and others to pollinate their crops and trees. This is an important aspect of today's apiary industry and will increase in importance in the future. Honey bees are by far the most important pollinators in the world, pollinating about 70 per cent of the types of food that we eat and 30 per cent of the actual food that we eat. Historically there has been a great reluctance by many commercial crop growers to pay for the rental of hives, but this attitude is slowly changing. In the past many growers have been able to rely on feral colonies of bees to pollinate their crops although as (or if) *Varroa* spreads across Australia, feral colonies will disappear and this benefit will quickly end.

It should be noted that providing hives for pollination purposes is a specialist activity as it requires a good level of knowledge and experience to have hives at optimum bee levels prior to placement, not to mention a lot of hard work for the beekeeper, both in preparing the colonies, delivering them and arranging to collect them again after a relatively short period. Providing hives for pollination may mean a reduced amount of honey being collected yearly and anyone wishing to offer this service needs to understand both the advantages of an attractive new source of revenue plus the disadvantages that providing this service may bring.

The preparation of colonies for pollination needs to be thought through and planned well in advance. A typical hive used for pollination consists of two brood boxes containing about a box and a half of bees. That is, the two brood boxes need to be almost full of bees for the hive to be acceptable to a crop grower. Also, since adult bees need to be about three weeks old before they start foraging for pollen there needs to be a high proportion of older adults in the colony. The colony will also need to have a high requirement for pollen and this is caused by the colony having a large number of uncapped brood that need feeding by the nursery bees.

To provide colonies that satisfy these requirements — large numbers of bees, many older adult bees that are foragers, and large numbers of uncapped brood that need feeding — takes careful planning and early preparation of the hive. If the pollination service is to be provided early in the spring, say for the almond pollination in August in north-east Victoria, preparation of the hives must start before winter and is likely to include supplementary feeding of both syrup and pollen substitute.

Hives destined to pollinate late-flowering crops may need to have frames of honey or capped brood removed to give the queen the room needed to continue to lay eggs.

Colonies to be used for pollination at any time of the year must have plenty of spare brood cells available so that the queen will not slow down her egg-laying rate. In addition, a congested hive may swarm which also needs careful management by the beekeeper to control.

Another unique feature of hives used for pollination is that the hives need to be moved often many hundreds of kilometres to the crops. Moving the hives is best done at night and the colony needs sufficient air to breathe. Most professional apiarists today shift their colonies 'open entrance', that is, they do not close the entrance and accept that about 10 per cent of their bees will be lost in transit. Professional beekeepers can save a lot of time by not closing the hive entrance and also know that the loss of about 10 per cent of their bees will be made up in about one week. Another popular method of moving hives long distances, particularly for hobby beekeepers, is to use a mesh cover instead of a lid to allow fresh air to circulate through the hive.

7.25

7.25: Providing hives for almond pollination in north-east Victoria.

Hives should be placed near the crop to be pollinated when about 10 per cent of the crop is in flower. Earlier than this and there may not be enough nectar or pollen for them to live off and they may visit neighbouring crops owned by other people to collect food. Pollinating a nearby crop owned by an adjoining landowner may be an issue for the farmer paying for the pollination service. The hives should be removed from the crop when about 90 per cent to 95 per cent of the crop has stopped flowering. Many new beekeepers see pollination services as a quick way to make ready money but, as mentioned, providing healthy hives in optimum condition at the right time is a skilled and time-consuming occupation. Brokers who manage pollination contracts employ specialists to check that bee cover in hives is satisfactory, as the farmer does not want to pay for colonies that are weak or unable to work to capacity. In large monoculture crop areas where bees are brought in from great distances by various beekeepers, the chances of the transfer of disease and pests is greatly increased and these will need to be dealt with or controlled when the hives are removed.

Another important part of managing hives for pollination is to ensure that your bees are not killed by poisonous sprays applied either to the crop to be pollinated or to other nearby crops. This is best done by talking with the crop owner about any plans to spray near your hives and also by talking with other farmers or crop sprayers locally about their activities. If you are in doubt about the effect of the spray or its possible drift towards your hives it is best to remove them from areas of risk. It is seldom that a beekeeper providing pollination services will be fully or even partly compensated for lost hives caused by poisoning.

COLLECTING POLLEN

Harvesting pollen is a good activity during the spring and summer when the flowers are blooming. Attaching a pollen trap to your hive is straightforward and you can collect a lot of pollen every day. In practice, since collecting pollen restricts how much pollen the bees are able to store and use to feed their own brood, I only use a pollen trap every other day. On the days that I am not collecting pollen I leave the entrance to the trap open so that the returning foragers can enter and leave the hive freely. Pollen absorbs water and will quickly grow mildew if it is not harvested every evening. Once I harvest the pollen I place it in a plastic bag and store it in the refrigerator or freezer.

SUMMARY

- Managing the colony to get it ready for spring and summer is an important activity that needs to be performed by the beekeeper. If you wish to have a strong colony of bees ready for the main nectar flow feed the bees a 1:1 sugar—water syrup mixture from about late August onwards. The first hive inspection of spring is important, not only because it will be the first time that you have inspected the colony for a number of months, but also because you need to take the hive apart and clean a number of its components, particularly the base.

- Hives should be inspected at least every three weeks during spring and summer. A good mnemonic to remember for what you need to look for when inspecting a hive is FEDSS — Food, Eggs, Disease, Space and Swarming.

- Colonies are likely to swarm during spring or summer and managing this is a legal requirement for beekeepers, particularly in urban areas. Two of the most popular methods of managing swarming are the split hive and Demaree methods.

- If you aim to provide hives for pollination make sure that the colonies are strong or the farmer may not accept your hives. Providing hives for pollination is a skilled occupation and needs careful planning. You may need to feed the colony a 1:1 sugar—water syrup well before the pollination contract to ensure that the queen lays to capacity and that the colony becomes strong.

8.

Autumn and winter management

AUTUMN

Autumn is a critical time for hive management as the colony needs to be protected against robbing by other bees and also needs to be shut down ready for winter. If either of these two activities is performed poorly the colony may not survive or may enter spring with insufficient numbers of healthy workers available to forage for the nectar and pollen necessary to build up brood numbers.

FEEDING READY FOR WINTER

A colony needs between four and six full frames of honey to survive over winter. A colony in the cool temperate conditions of Victoria and Tasmania will need greater stores than one in the tropical north that does not need to use a lot of energy to keep warm. When starting your beekeeping activities it is better to err on the side of the bees and leave them at least six full frames. Many southern hobbyist beekeepers feel comfortable leaving a full super of eight frames for their bees over winter. Do not over rob your bees as you will end up with a weak colony going into winter and one that may not survive to see the spring. If sufficient numbers of full frames of honey are unavailable, either due to a poor season or because the beekeeper has robbed the hive of too much honey late in the season, the colony will need to be fed syrup that it can

8.1: An experienced beekeeper shows beginners how to shut down a hive for winter.

8.2: Cleaning a hive.

8.3: Apiary ready for winter.

use to build up its stores. The previous chapter on spring and summer management explains the main methods used to feed sugar syrup and pollen/pollen substitute to bees (see pp. 120–4). The use of sugar syrup is fine while temperatures are high enough so that the colony has not formed a winter cluster. If the colony runs out of honey during the winter either add a frame of stored honey, candy or sugar syrup close to the cluster so that the colony will be able to easily reach it without overly breaking the cluster. In some of the warmer areas of Australia winter temperatures may be sufficiently high that the colony does not form a cluster and there may even be forage available at some locations over the winter months.

Candied sugar, though, is a last resort for the bees and cannot be relied upon to provide all the carbohydrates the bees require over a long winter since candied sugar is not that easy for the bees to consume. White sugar from the supermarket contains too little water for the bees to easily eat; candied sugar contains over half its volume as water, which the bees can more easily consume.

MINIMISING ROBBING

Most robbing of colonies occurs when nectar supplies are scarce and other honey bee colonies are desperately looking for food. In Victoria and Tasmania this typically occurs during the autumn. During this time the European wasp is often out in strength and these too can rob the hive of both honey, brood and adult bees. In other parts of the country, such as the Sydney region, one of the biggest honey flows is during the late autumn and thus robbing is less of a problem there at this time of year. To restrict robbing you should:

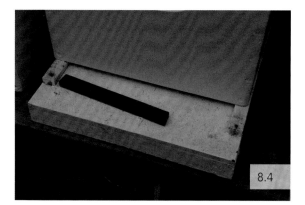

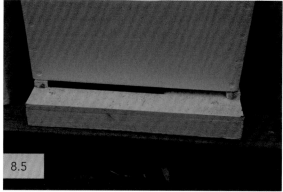

8.4: Many beekeepers reduce the size of the entrance before winter to minimise the amount of cold air entering the hive.

8.5: The entrance reducer inserted into the entrance of the hive.

8.6: Bees feeding on pollen substitute.

- Reduce the entrance of the hive to a width of 5 to 10 centimetres. Bees are better able to defend the smaller entrance from intruders.
- Ensure that when you inspect the hive it is left open for the minimum amount of time.
- Do not leave burr comb, frames of honey or spilt syrup near the hive to attract robbers.

When you are near your hives, keep watching the entrance for unusual activity. Robber bees are very fast and aggressive in their attempts to enter the hive. If you notice this type of activity shut down the entrance to the hive completely for an hour and the robber bees may go away.

Not long ago I inspected one of my hives on a Sunday and found that it was full of honey. It was too late in the day to extract so I planned to return the following weekend to remove the full frames. I returned the following Sunday and, to my astonishment, all the frames in the super were completely empty. Robber bees had found the colony and had stolen all of the honey, leaving an empty super.

Apart from mass robbing occasionally occurring, robbing attempts by individual intruders is an everyday occurrence and often just watching a hive entrance you will see bees pushing other bees away. This pushing away seldom involves stinging but the intruder soon gets the message and moves on. Potential robbers being pushed away from the hive often have different body markings and colours to the colony they are trying to penetrate.

THE WINTER SHUTDOWN

There are four reasons why, after the summer is over, a hive needs to be prepared for winter.

- The hive probably consists of between three and four boxes and this is too many for the much smaller number of bees to keep warm over winter and to protect against pests such as Wax Moth and mice.
- Depending on the region of Australia, the beekeeper needs to ensure that the colony has sufficient honey and pollen available to survive over winter. This, sadly, is often not the case and the main reason that previously healthy hives do not survive winter is that they run out of honey and are unable to keep the colony warm enough to survive. Bees around the Sydney area of New South Wales rarely run out of honey over winter because of Melaleuca trees and successful late autumn foraging.
- Before the start of winter, pollen or pollen substitute may need to be provided by the beekeeper, so that if pollen supplies are short during early spring the workers can use the pollen supplement to make royal jelly to feed to the developing larvae.
- Check on the strength and health of the colony and confirm that there is a queen present.

The climate in the region of Australia in which you live and the trees that are in flower during autumn will influence the date of your winter shutdown. Towards the end of April to late May you will need to take the following steps to prepare your hive for winter.

On a warm autumn day reduce the size of the hive down to one or two boxes. Assuming that you have a three-box hive, consisting of two brood boxes and one super, in order to reduce the number of brood boxes down to one or two; do the following:

- First remove the lid from the hive and also place the top box, or super, at the side of the hive.
- Next remove the queen excluder and place that at the side as well.

- Place a plastic container that can hold the empty frames next to the hive.
- Remove each full frame from the super and shake or brush the bees back into the hive.
- Now place the frame, without any bees attached, into the plastic container and shut the lid — if you do not shut the lid, bees from the hive will smell the honey and within a couple of minutes the frame will again be covered with bees.
- Take the second frame out of the super and repeat the procedure until all of the frames have been removed from the super and placed in the sealed plastic container.
- Alternatively, a clearer board can be used to remove most of the bees from the super down to the brood boxes, as explained in the chapter on harvesting honey.
- Take away both the empty super and the box of frames.

I used to reduce my brood area down to one box over winter; more recently I only reduce it to two boxes. As long as there is not a queen excluder separating the two boxes the bees will be happy in either one or two boxes over winter.

The main exception to the above procedure occurs if you need to reduce the number of brood boxes, say if you have a three-box hive and do not use a queen excluder, and you do not want to kill any brood in the top brood box. Reducing the number of boxes in this situation is more time consuming and needs to be performed over a few weeks. Remove the top brood box and place it above a queen excluder making sure that all of the bees have first been shaken off into the bottom brood box. This will ensure that the queen is in the bottom brood box. When the brood box is placed above the queen excluder, worker or nursery bees will quickly move through the excluder to look after the brood left behind, leaving the queen below the excluder. Return to the hive after three weeks. All of the brood should have hatched in the top box and the box can now be removed as explained above.

If the frames are only half full of honey and you do not want to store them, use the following method to remove the honey from the frames before taking them away for storage. Remove each frame in turn, score the surface of the capped honeycomb with a scratcher and return the frame to the super. Repeat this process until all of the capped honeycomb in the super has been scored and honey is flowing freely from the cells. Carefully turn the super upside down and place it back onto the hive. Since the cells are now facing downwards all of the honey will flow out and will be quickly cleaned up by

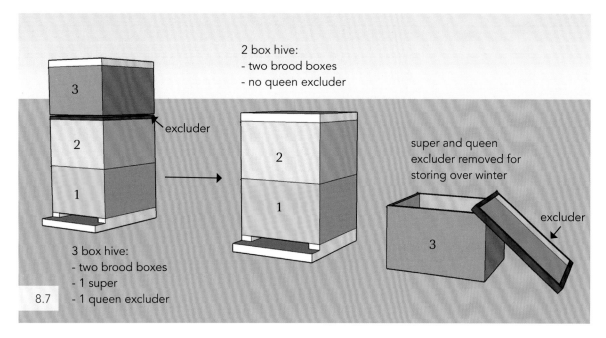

2 box hive:
- two brood boxes
- no queen excluder

excluder

super and queen
excluder removed for
storing over winter

excluder

3 box hive:
- two brood boxes
- 1 super
8.7 - 1 queen excluder

8.8

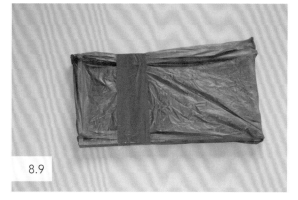

8.9

the bees. The bees will now store the honey in the lower parts of the hive since they will be unable to replace it in the downward-facing cells.

After a few days the frames will be clean and can be removed and stored. Turning a super upside down after first scoring the capped honeycomb is also a good way to remove hard or crystallised honey from a frame. The use of a queen excluder to stop the frames falling out when the super is turned upside down helps a lot.

8.7: Reducing size of hive for the winter. Both a super and the queen excluder have been removed from the three-brood box hive, leaving a two-brood box hive.

8.8: I use an airtight plastic storage bin to store frames of honey over winter. Note that I have attached two pieces of wood at either side of the container to support the frames.

8.9: Full frames of honey can be stored in polythene bags from the supermarket. Before placing the frames in the plastic container I freeze the frames in the freezer for two days to kill Wax Moth or other larvae.

8.10

8.11

CHECK THAT THE HIVE HAS SUFFICIENT FOOD TO LAST OVER WINTER

Before placing the lid back on the hive, check that there is sufficient honey left in the brood boxes, about four to six full frames, for the bees to live on over winter. If not, and you have some full frames of honey taken previously from the same hive or some frames from other hives that are disease free, replace some of the emptier frames in the hive with full ones. Previously stored honey frames are best because honey contains minerals and enzymes that sugar syrup does not. If you do not have any full frames of honey available, follow one of the syrup-feeding procedures described earlier in Chapter Seven on spring and summer management. Before winter, if syrup needs to be fed to the colony so that they can quickly convert this into honey to store, use a ratio of 2:1 sugar–water by weight for the syrup. The bees can much more easily convert

8.12

8.10: Removing half-full frames of honey. Step 1: If you have half-full frames of honey that need to be removed for the winter, one way of getting the bees to remove the honey for you is to score the surface of the capped and uncapped comb. Next, return the scored frames back to their super before placing the super upside down on the hive. This is also a good method for removing hard candied honey from frames.

8.11: Step 2: Turn upside down the super of scored frames and place it upside down on the hive. A queen excluder is used to stop the frames from falling out when you are turning the super upside down.

8.12: Step 3: The last step is to remove the queen excluder before placing a lid on the hive. Within a few days the bees will have removed all of the honey and stored it elsewhere in the hive. The empty frames can now be removed and stored for the winter.

this mixture to honey than the 1:1 ratio sugar—water mixture used in the spring to encourage the queen to lay eggs.

Most experienced beekeepers leave four or five full frames of honey in the hive over winter for the colony to feed on, and up to six frames if the hive is kept in a particularly cold location such as parts of Victoria or Tasmania. If too many frames of honey have been removed during the autumn, the colony may not have sufficient stores of honey to survive the winter. At the start of spring, finding no live bees in your hives and only dead ones with their heads pushed deep into cells as they desperately sought food is particularly distressing and a sign of very poor management.

CHECK THE STRENGTH OF THE COLONY

While you are inspecting the colony check how many bees are present. Going into winter with a weak colony will mean that the colony will not be able to build up strength by the time the first honey flow starts in early spring. If the colony is weak at the start of the winter and you are unable to feed it sufficiently so that it will increase in strength there is no point taking the weak hive through winter and then having to deal with it in the spring. Culling weak or diseased hives is best done during the autumn. It is the time to merge any weak hives with stronger colonies.

MAKE SURE THAT THERE IS A LAYING QUEEN PRESENT

Check the hive for a laying queen. If you are unable to locate the queen look for eggs or young larvae and, if these are not present, you may need to either re-queen or merge the colony into another hive for the winter. If only capped brood is present, however, it may be impossible to introduce a new queen to save the colony.

If brood is present, look for an even brood pattern across a frame. Patchy brood may indicate a weak or failing queen or even disease.

If you decide that you need to re-queen you will have to do this earlier in the autumn. A young, vigorous queen will continue to lay later into the autumn and will also resume laying earlier in the spring ready for the first honey flow. Colonies headed by young queens are also much less likely to swarm during their first spring.

Even if there is a queen present she may not be laying eggs at that moment. Check again in a couple of weeks to confirm if she is present or not.

MAKE SURE THAT THE COLONY IS FREE FROM DISEASE

Check for signs of disease as this will need to be treated or managed if the colony is to survive and be strong, ready for the spring nectar flow. If the hive contains Small Hive Beetle you will need to introduce beetle traps or an approved miticide trap to kill the beetles.

A strong foul smell may indicate that American Foul Brood or European Foul Brood is present. Treatment for these two diseases is discussed in Chapter Fifteen on brood diseases.

OTHER THINGS TO CHECK FOR DURING AUTUMN MANAGEMENT AND PREPARING FOR WINTER

Checking the hive during autumn is crucial to making sure your bees and the hive are healthy and will function well in spring. Some other tasks to do at this time are listed below.

- While you are inspecting the hive check for any damaged parts of the hive that will need replacing or repairing over winter. Remove any damaged parts of the hive and ensure that the hive is in sound condition.

- You may want to reduce the size of the entrance to about 5 to 10 centimetres to protect against robbing. Many beekeepers believe that this is useful to protect the colony against cold air although my experience is that a hive that has appropriate levels of honey in reserve does not need protecting against the cold. Also, if mice are a problem in your area consider adding a mouse excluder to the entrance to stop them entering the hive and making a cosy home for the winter, destroying comb in the process.
- In many parts of the country warm winters may mean that the colony can remain active all year long and there may not be a need to close down the hive over winter.
- Confirm that the hives are placed in a suitable location, free from drafts and facing the winter sun. Provide a wind break if necessary. Also, raise the hives 30 to 40 centimetres above the ground away from dampness; this will also allow fresh air to circulate.
- Remove any frames that need replacing due to damage or dark or poorly formed comb.
- If necessary, rearrange the frames so that brood is in the centre of the hive with stores of honey and pollen next to them.
- Control weed and grass growth around the hive and keep the general area of the apiary clean of rubbish.
- If large animals are present, either fence off hives or use Emlocks to keep them closed in case of a disturbance. Do not keep the Emlock strapping really tight as it may break, particularly after a lot of use or in cold weather.
- Some beekeepers completely strip down the hive in the autumn, rather than in the spring, to replace damaged equipment and to clean the boxes, lid and base.

WINTER

Depending on the region of Australia that you live in, particularly in parts of Tasmania, the mountainous regions of Victoria, southern New South Wales and the Australian Capital Territory, you may not be able to open your hive between early May and late August. Much depends on the temperature since exposing the colony to cold may quickly harm the brood and severely reduce the strength of the hive coming into the spring honey flow.

Honey bees do not hibernate over winter but, like penguins, cluster together to keep warm. The colder the air is around them, the tighter the cluster. The bees on the outside are closest to the cold and regularly change positions with bees deeper in the cluster in order to keep themselves warm. When the air temperature inside the hive drops to 18° Celsius the bees start to form a loose cluster, and when the hive temperature has dropped to 14° Celsius all of the bees are included in the cluster. It is important during the winter that this cluster is not broken during a hive inspection as many brood may die as a result. In all but the very coldest parts of Australia hives will carry brood throughout winter.

The adult bees generate heat by shivering, but there are limits to how much heat this can produce. As the temperature drops the cluster contracts into a tighter ball and it is this tighter ball that keeps the brood warm.

There may be drone cells left late in the autumn or even during the entire winter in the northerly, warmer regions of Australia. Drone cells are usually located along the bottom edge of a brood frame. If the temperature drops and the workers form a cluster around the brood, as the cluster grows smaller, fewer of the drones will be covered and so may die. This is another way of nature saying that male drones are dispensable when conditions are poor and that the focus of the colony is to maintain female brood since they will ultimately be needed to look after the colony.

WINTER MANAGEMENT

Winter management of the hive comes down to ensuring that there is sufficient honey available for the colony to survive on until spring. Since poor weather may make it impossible to open the hive to check how much honey is left, the best method of checking honey stores is to lift one end of the hive to see how heavy it is. A heavy hive has plenty of honey available for food while a light hive has little honey available for food and the colony needs feeding immediately. This can best be done by quickly taking the lid off the hive and replacing an empty frame with a frame full of honey stored since the previous autumn. Make sure though that the new frame of honey is placed next to the cluster since if the honey is too far away in the brood box the workers will not be able to break the cluster during very cold conditions to reach it and may starve to death anyway.

If a spare full frame of honey is not available, another way to feed emergency sugar to a starving colony is to place a block of candied sugar above the hive mat as described on p. 123.

SUMMARY

- Preparing your hive for winter is an important activity and needs to be performed by the beekeeper during autumn. If the hive includes a super and a queen excluder, both need to be removed leaving only one or two brood boxes for the colony to overwinter in. Winter survival of the colony is dependent on there being sufficient stores of honey present for the bees to feed on. In the colder regions of Australia six full frames of honey will need to be left for winter, and a smaller number in the warmer regions or where there is still a good honey flow during the colder months.

- Robbing may become a problem during the autumn as there is generally little forage around and nearby hungry colonies may target your hives if they offer ample supplies of honey. One way to minimise robbing is to close down the size of the hive entrance to about 10 centimetres.

- Once the temperature outside drops below 17° Celsius the hive should not be opened for long periods and brood frames should certainly not be removed for inspection.

- Take the opportunity presented by the autumn closure of the hive to tidy up around the hive area, to make any repairs and to think about any new boxes and frames you will need to make over the quieter winter period.

Extracting

The first extraction for a new beekeeper is always both exciting and daunting. Exciting because all the hard work, time and effort by the bees and beekeeper has finally come to fruition. Daunting, as extraction often appears to be a difficult task, which it is not, and the process raises the added concerns about how much honey will be extracted and its quality. My experience with new beekeepers indicates that they usually underestimate how much honey a hive will produce. The amount obviously varies according to the season, the available forage and, of course, the weather. In a good year roughly 20 to 30 kilograms per extraction is obtained and this may be repeated approximately two or three times each season. This amount of honey can be obtained from approximately ten or twelve full frames of honey. In favourable conditions the average hive will provide about 50 to 80 kilograms per season to the hobby beekeeper. This is a lot of honey.

Many new beekeepers believe that the only way to extract honey is to use an extractor. While it is true that either a hand-cranked or electric extractor makes harvesting of honey easier, there are other inexpensive ways to do the job. I will begin with a discussion of the inexpensive labour-intensive ways to extract honey and will go on to discuss the use of an extractor at the end of the chapter.

USING A SIEVE TO HARVEST HONEY

If you do not have access to an extractor, the use of a sieve or strainer to drip and filter cut out comb is an inexpensive and simple way to harvest your honey. Begin by cutting the capped comb out of the frame in the area immediately under the first frame

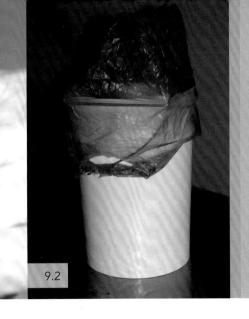

9.2

wire. Leave the comb above the wire, where it joins the top bar, untouched. This will provide the bees with a straight edge to redraw the comb when the frame is returned to the hive. Place the cut-out comb into a large sieve, chop it up with a knife and let the honey slowly drip through the mesh into a bucket or other container. As the honey needs to be liquid for this type of extraction to be successful it is better to choose a warm day. Alternatively leave the sieve and bucket in a warm kitchen to drain slowly. As the honey drains from the comb, periodically chop up the comb and compress it to allow more honey to escape. A quick way to speed up extraction using this method is to leave the sieve and bucket for a short period in filtered sun where the heat will quickly liquefy the honey and it will soon drain into the bucket. A plastic bag will need to be placed over the sieve to stop flies and dust contaminating the honey. Also keep watch for bees that will appear very quickly to rob the harvest if the sieve and honey are not completely sealed. It is important that you do not leave the honeycomb in the sun for too long or the honey will overheat and turn brown, losing much of its flavour and goodness.

When removing sections of comb out of frames, cut half a frame length at a time between a pair of wires. Longer lengths are difficult to handle and are likely to fall outside the bucket or container holding the sieve and create a mess.

A simple device to make and use when cutting comb out of frames is a length of wood 2.5 centimetres by 5 centimetres with a screw inserted in the centre of the long side. Choose a screw that is long enough to protrude through the 2.5-centimetre depth of the wood by at least 1 centimetre. The device is placed horizontally across a bucket with the screw point facing upwards. When the side of the frame is balanced on the screw it can be swivelled easily and the comb, as it is cut out, will fall into

9.3: When you cut out sections of comb for placing in a sieve leave a strip of intact comb at the top for the bees to use as a starter to build new comb. This frame was left overnight in a hive and the colony had already started to build new comb from the starter strip.

9.4: Using a scratcher to uncap honey comb is an easy and inexpensive way of preparing frames for the extractor.

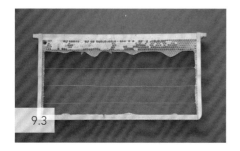

9.3

9.4

the bucket below. After this the cut-out comb can easily be moved from the bucket to the sieve without creating a mess.

As previously mentioned, make sure that when you cut out the honeycomb for extracting that you leave a strip of honeycomb above the top wire. The bees will use this as a template or starter to build new comb and you will not need to attach new foundation to the frame. Bees take longer to make their own foundation and also consume additional honey to make the wax. The benefit though is a more natural cell size and cleaner comb for the bees to use.

A frame is generally ready to harvest when it is more than two-thirds full of capped honey. If in doubt turn the frame on its side and give it a shake — if any runny liquid falls from the frame it is probably not ready to harvest as the watery honey may ferment, spoiling the rest of the harvest.

USE OF AN EXTRACTOR TO HARVEST HONEY

If you have access to an extractor the uncapping process is different since you will leave most of the comb intact on the frame and remove only the surface capping that has been used to seal the ripe honey. To remove the capping the same wooden support and bucket described earlier is used. Again, a frame is ready to harvest when it is more than two-thirds full of capped honey.

REMOVING CAPPINGS

METHOD 1

Holding a capped frame over an uncapping bucket, use a scratcher to score or break up the capping. Once the surface has been broken up, repeat the process on the other side of the frame. The frame should now be placed inside the extractor. Repeat the process on the remaining frames until the extractor is full of uncapped frames.

A scratcher, although easy and flexible to use, has the disadvantage that it produces a lot of wax particles that will soon block the sieve used to filter the honey. As a result the sieve will need to be cleaned frequently during the extracting process.

9.6

9.7

9.5: Using a knife to uncap honeycomb. This is slightly more difficult than using a scratcher but the advantage of it is that the capping comes off in sheets and does not block the honey filter.

9.6: A convenient way to uncap frames or to remove sections of comb for filtering in a sieve is to use a wooden bar with a screw or nail pushed through it. The wooden bar sits on top of a plastic bucket.

9.7: Before using knives to uncap frames of honey heat them in hot water so that they will more easily cut through the wax.

9.5

METHOD 2

An alternative to using a scratcher is to use a long knife to cut off the capping. As before, the frame is rested on the screw and, with a downward sawing motion, the sharp edge of the knife slices the capping off the comb. Using a knife is slightly more difficult than using a scratcher, but has the advantage of peeling off the capping in a single sheet that does not block the filter with fine wax particles.

Simple extracting knives are inexpensive and may need to be heated in hot water before use. Many beekeepers use two knives, one of which is heated in hot water while the other is being used to cut the capping. When the knife used for extracting cools down, it is returned to the hot water and the other knife removed, dried and used. On warm days, when I do most of my extracting, I have often used a knife without first heating it since I find the wax capping is sufficiently warm and pliable to be cut without the need to heat the knife. If you are heating a knife using water make sure you dry it quickly so that you do not contaminate either the frame you are uncapping or the cappings in the bucket below with water dripping off the knife.

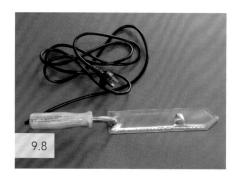

9.8

9.8: An electrically heated uncapping knife is a convenient way to remove cappings from honeycomb.

METHOD 3

The most convenient way to remove cappings is to purchase an electrically heated knife that will easily cut through the wax capping. Electric knives have a heating element and thermostat inside the blade. Some have adjustable thermostats, but most are factory set and will remain at the correct temperature for cutting through wax during uncapping. The knife is moved fairly quickly through the surface cappings in order to avoid scorching the honey contained in the uncapped cells.

Steam-heated knives used to be popular and remain available at most beekeeping supply shops. Steam knives are cheaper than electric knives, but have the major disadvantage that they require a water boiler, which the beekeeping supply store is unlikely to provide. Steam knives are also more cumbersome to use since there is a pair of hoses connected to the knife, the first to provide steam to heat the knife, the second to move the used steam away from the beekeeper. Some older beekeepers prefer to use a steam knife as they feel a steam knife retains its heat better than an electric knife, even though it is less convenient to use because of the need for an attached water boiler and steam pipes. If a beekeeper has simple plumbing skills a pressure cooker can be used to make a steam boiler. It is imperative when producing steam under any pressure that the boiler is not allowed to burn dry.

USING AN EXTRACTOR

Once the capping has been scored or removed it is time to place the frames into an extractor to harvest the honey. In the commonly used tangential extractor, frames can be inserted so that either the top bar or the bottom bar rotates first, and in either direction honey will flow out.

As a general rule, when determining the number of turns that the frame should be rotated the 75–150–75 rule applies. First, rotate one side of the frame 75 turns to remove half of the honey from one side. Turn the frame over and rotate the next side for 150 turns to remove all of the honey from this side. Now return the frame back to its initial position and rotate a further 75 turns. This process is used in order to prevent the frame 'blowing out'. If for example you start by rotating the first side of the full frame for 150 turns the weight of honey on the unspun side remains the same whereas the side that is being spun is becoming more fragile and loses substance and strength as the honey is thrown from the comb.

The unspun side of the frame, with the centrifugal force of the extractor and its weight, may push through the foundation, breaking the honeycomb and resulting in a

9.9

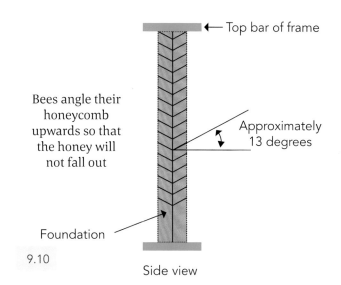

Top bar of frame

Bees angle their honeycomb upwards so that the honey will not fall out

Approximately 13 degrees

Foundation

9.10

Side view

9.9: Comb does not point out horizontally from the frame but slopes upwards at about 13 degrees. This needs to be taken into account in a radial extractor but not in a tangential extractor, commonly used by hobby beekeepers. In a tangential extractor, frames can be inserted so that either the top bar or the bottom bar rotates first.

9.10: Worker bees build comb so that it is sloping slightly upwards. This minimises the chances of either honey or brood falling out.

9.11: In a tangential extractor commonly used by hobby beekeepers, frames can be inserted so they rotate either top bar first or bottom bar first.

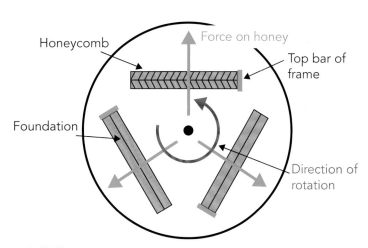

Honeycomb

Force on honey

Top bar of frame

Foundation

Direction of rotation

9.11   Top view of a three-frame extractor

lot of extra work for the beekeeper to repair or replace the wax foundation.

Sometimes, with thinner foundation, the comb will break during extraction. This is not a problem to the bees; just use your fingers to repair the comb so that there are no wide splits in it, return it to the hive and the bees will soon repair the damage leaving a flawless comb to store honey in.

If you have, say, only two frames left to extract and are using a three-frame extractor, you will need to keep an already extracted frame in the extractor cage so that the unit does not become unbalanced and wobble when rotating. When the two frames are empty of honey, you can remove all three frames and the extraction is over.

After the frame has been removed from the extractor it is called a 'sticky' since it is covered in honey and sticky. If you do not plan to return the sticky frames to the hive after extraction, they cannot be stored covered with honey since this will attract bees, ants and other insects. The frames need to be returned to the hive for 24 hours immediately after extracting so that the bees can clean them, removing all the honey remnants, leaving the frames and comb spotless of honey and wax particles. After 24 hours you can remove the clean frames permanently from the hive for storage if required with little risk of attracting Wax Moth or Small Hive Beetle, unless the frames contain large amounts of protein from larvae skins or pollen, which Wax Moth larvae like to eat.

One point worth stressing is the necessity to uncap frames and to extract in a room, shed or garage that has doors that can be closed tight to keep out inquisitive bees. It only takes one or two bees to smell the honey and they will soon return to the hive to tell their sisters about the bountiful supply of delicious food that is currently waiting for them in your chosen extracting space. Very soon this space will be full of hundreds of bees all looking for the honey that you are extracting. Many years ago when I was new to both bees and extracting I set up my extractor in the garage, spun a few frames, and went inside for a cup of coffee, forgetting to shut the garage door. When I returned about twenty minutes later there were literally thousands of bees in my garage not only covering the extractor, frames and uncapping bucket, but flying around and covering just about everything else. Removing the bees was almost impossible and I had to patiently wait until dusk, when they returned to their hives. Only then could I continue extracting.

In addition to the above methods of honey extraction see Chapter Twelve on sustainable beekeeping on the use of a fruit press to extract honey.

CHOOSING AN EXTRACTOR

There are a variety of designs of extractor on the market and the type of extractor that you buy will depend a lot on individual preference. There are, however, some general points that you need to consider when making your choice.

- First, is the extractor made from food-grade stainless steel? This type of steel is non-magnetic, so a simple test to confirm if the steel is food grade is to touch the extractor with a magnet and see if the magnet is attracted to the metal. If it is not attracted you can be confident that the extractor is made from food-grade stainless steel. It is worth noting, however, that some parts of the extractor may not be food grade — for example, the metal gearing. This is not a problem as the gears will not usually come in contact with the honey being extracted.

- Does the gate, at the bottom of the extractor to release the honey, leak when closed? This may be difficult to check, but look to see if the valve shuts tightly. In addition try opening and closing the gate several times to see if the bolt holding the moveable lid of the gate becomes loose. A further point to check is that the rubber or plastic O-ring sealing the gate entrance protrudes a reasonable distance above the body of the metal or plastic gate, say by 1-millimetre or more.

- If you plan to extract from Ideal frames, make sure that they will be held in the extractor cage and not fall through to the bottom of the tank. If the extractor is in all other respects a good buy, this is not difficult to fix. Some stainless steel frame wire attached across the base of each frame-holding cage will generally prevent the frame from falling through the base of the cage and into the tank. If you are spending a lot of money on a new extractor and use large numbers of Ideal frames you may be better buying an extractor that has a cage designed to support both full depth and the Ideal size of frame.

- Check that the frame, when it sits in the cage, does not touch the cross bar of the extractor when it is rotated, hindering rotation of the cages.

- Consider whether the honey can escape freely from the tank through the honey gate without building up and touching the bottom of the cage. Honey that has built up and touches the bottom of the cage will stop the cage from being turned and put undue strain on the gearing mechanism.

- Can the cage be easily removed for cleaning?

- Ask if the spindle at the centre of the cage sits on a ball bearing at the bottom of the tank. If it does this will make turning the cage of the extractor much smoother and easier. You should also check if replacement ball bearings are available should you lose the one in the extractor while cleaning it.

- What is the warranty period and are spare parts readily available?

- Are the gears made of iron or nylon? Nylon gears are light but will wear out quickly and replacements may not be cheap or easy to obtain a few years after you purchased the extractor.

- Many older second-hand extractors are made of galvanised steel. Galvanised steel is corroded by honey and is today not used to make extractors for reasons of food hygiene. Take this into account when offered an apparent second-hand bargain.

- Some extractors have a plastic drum and are cheaper than extractors with stainless steel drums. I have used a plastic extractor several times and, although I would prefer to use a stainless steel extractor, plastic extractors are cheaper and are durable if handled with care.
- The handle on the extractor may be mounted on top or on the side. Top-mounted handles may be less expensive than side-mounted handles, but the difference in cost is not significant. Try both types of handle and decide which you prefer before choosing one or the other.

Electric extractors are much easier to use than manual extractors, but are significantly more expensive. The motor on the extractor needs to be powerful enough to turn a cage containing full frames without overheating. Electric extractors are a good option if you can justify the cost as they take all of the hard physical work out of extracting. Check to see if there is a switch to rotate the motor in either direction, and that the motor includes a variable speed control so that you can adjust the speed of the cage rotation.

For safety reasons the extractor needs to be earthed and confirmation obtained that it is made to a recognised international electrical safety standard such as for the European Union. Electric extractors made in China usually have motors that operate on 220V, not the Australian standard 240V. Watch out for this as the voltage difference may result in the motor overheating in use, particularly at higher speeds or with heavier frame loads. Transformers are available that convert the 240V power used in Australia to the 220V power that Chinese motors are designed to use, although these will add at least another $200 to the cost of the extractor.

9.12: Electric extractors are much easier to use than manual extractors.

9.12

TYPE OF EXTRACTOR TO BUY

Extractors come in two types: tangential extractors and radial extractors.

TANGENTIAL EXTRACTORS

Most hobbyists use a tangential extractor due to their lower cost and also because they take up much less room. As the name suggests, the frames in a tangential extractor lie tangentially to the direction of rotation of the cage. Tangential extractors are compact, inexpensive and are easy to use and clean. They also keep their resale value and are a good investment, particularly if the cost is shared by a few beekeepers.

Tangential extractors can usually hold two, three or four frames at a time. My experience is that although a two-frame extractor sounds like a good deal due to its lower cost and compact size, they are not very practical since the frames are placed too close to the spindle, which is the axis of rotation of the cage. As a result, insufficient centrifugal force develops to effectively remove the honey from the centre of the frame. This may result in a frame that has had its honey removed on either side, but not in the centre. This is both frustrating and a waste of time.

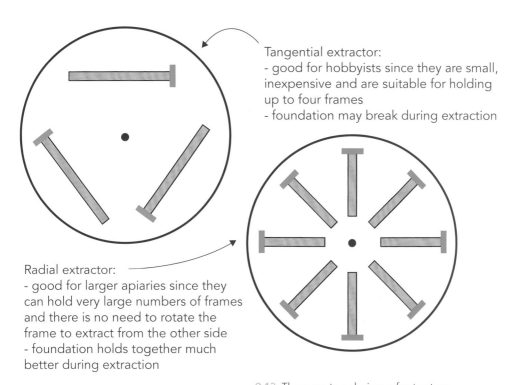

Tangential extractor:
- good for hobbyists since they are small, inexpensive and are suitable for holding up to four frames
- foundation may break during extraction

Radial extractor:
- good for larger apiaries since they can hold very large numbers of frames and there is no need to rotate the frame to extract from the other side
- foundation holds together much better during extraction

9.13

9.13: There are two designs of extractors available, tangential and radial. The tangential extractor is more widely used by hobby beekeepers due to its lower price.

RADIAL EXTRACTORS

Radial extractors are much larger than tangential extractors and are used almost exclusively by professional apiarists. Radial extractors can hold a large number of frames ranging from, say, six up to 40 frames; some very large ones hold even more. Unlike the tangential extractor, frames are placed into a radial extractor in the same orientation as the spokes in a bicycle wheel. Apart from being able to hold many more frames, due to the orientation of the frame a radial extractor has the added advantage that the frames in it can be rotated at much higher speeds without damaging the wax foundation. The much higher speed of rotation also leads to higher centrifugal force, enabling the honey to be spun from each frame much more quickly.

FILTERS

Some extractors have a built-in filter and holding tank. If you only have a small number of hives these are the most convenient to use since the honey is extracted and filtered in a single operation and is ready to use immediately after extraction. Manual extractors of this type are useful for beekeepers with up to ten hives. The only disadvantage of this type of arrangement is that the extractor tank must be cleared of honey before the next extraction is undertaken. This may be a minor inconvenience or can be overcome by buying a second tank for the extractor at the time of the initial purchase.

Many extractors do not have a built-in filter; instead, the honey, containing a lot of wax particles, pours out of a gate at the bottom of the extractor tank and into a pail resting on the floor beside the unit. In order to filter the honey, a sieve or other filter may be placed on top of the pail and as the honey flows from the extractor tank gate into the sieve it is filtered before passing into the pail. The sieve must be periodically checked and cleared of any build-up of wax or it will not filter efficiently. This arrangement has the advantage that when one bucket is full, another can be placed under the open honey gate and extracting can continue. This type of extractor is convenient for beekeepers with approximately six or more hives.

STORING FRAMES OF HONEY

As I described in Chapter Eight on autumn and winter management, if you only have one or two frames of capped honey and do not plan to extract for some time, you need to store the capped frames. To do this you must first wrap them in a new, large garbage bag. Seal the bag with thick tape. Then place the frames in the freezer for two days to kill any Wax Moth or Small Hive Beetle larvae. Unless you are lucky enough to have plenty of freezer space this will usually involve freezing one frame at a time. Store the frames, still in their sealed garbage bags, inside an airtight plastic bin until you are ready to extract.

9.14: Some extractors have a built in sieve that makes extracting and filtering much easier. This three-frame extractor includes a sieve and a tank at the bottom. Honey flows from the extractor sitting on top, through holes in the base, and into the sieve at the bottom where any wax particles are filtered out.

9.15: The bottom tank of the extractor includes a removable sieve that cleans the honey of any particles such as wax.

COMB HONEY

PREPARING FRAMES FOR COMB HONEY

Plastic 'sections' are available that are usually either square or round and, together with their plastic support, fit neatly into a frame. The bees draw comb and deposit honey within the section, which is later removed and sold. These sections of honey are expensive to produce and many colonies of bees will simply not draw them out satisfactorily. These days cut comb is the most common way honeycomb is sold, either in large square or round pieces cut from the frame or in small chunks of comb added to jars of liquid honey.

Begin your attempts to obtain your own cut comb honey by using shallow or Ideal supers. Use very thin, new, pale beeswax foundation and just add a small starter strip to the top bar of the frame and let the bees make their own thin, perfectly white, honeycomb. The shallow super used for comb should be placed on the top of the hive, under the lid in order to avoid unnecessary bee traffic that will discolour the comb surface.

Colonies of bees used to produce cut comb honey should be on a good honey flow for the best results. The hive will need to be monitored and the honeycomb super removed once the frames are capped. Small Hive Beetle prevention is paramount and the frames should be carefully wrapped in plastic and placed in a freezer for at least 48 hours to kill any larvae or eggs. Once removed from the freezer they should be left in the plastic to come up to room temperature to avoid any condensation problems.

EXTRACTING COMB HONEY

After defrosting and ensuring the frame of honeycomb is at room temperature, place the frame on to a kitchen cutting board and carefully cut the rectangle of comb away from the wooden frame surround. Prior to doing this, inspect both sides of the frame and decide which is the more visually pleasing side to display in a container. This side should be placed face-up on your cutting board. Using a sharp knife, you can carefully cut squares of honey and place them into a plastic box or container. Your knife can be heated to make the process a little easier and improve the clean cut of the comb. An alternative way to cut out comb honey is to use a large pastry cutter available from most kitchen supply stores. These are often round in shape. It is also possible to buy a large, square, specialised comb cutter from some beekeeping supply stores.

A piece of chunk comb honey is an attractive feature in a glass jar of liquid honey. Glass jars should be thoroughly sterilised before use. Once the chosen piece of cut comb has been removed, gently place it in the jar and slowly cover it with honey, ensuring that the honey is poured in gently enough so that air bubbles are not trapped spoiling the appearance of the product. If you have previously cut rounds of honeycomb for sale, the valuable leftover pieces can be used in this way.

For larger, square-cut or round-cut pieces of honeycomb I find that the least expensive and most useful containers are plastic takeaway food containers available in bulk from most discount or $2 type stores. If you require larger numbers of this type of container, enquire at some of your local takeaway restaurants where they obtain theirs and buy them directly from the same wholesaler.

REUSE OF WAX FOUNDATION

Most beekeepers will have wax left over after extracting and by the end of the year this can be a sizable amount. The question often asked is what can I do with the wax and how can I clean it? Cappings left after extracting or comb cut out of frames contain a lot of impurities: pollen and honey are present as are silk cocoons and other remains left by the pupae that need to be filtered out together with pieces of dead bee. The first stage in cleaning the wax is to melt it and then to filter it through an old sheet or piece of hessian sacking. The two most common methods used by hobbyists to clean wax are the saucepan method and the solar beeswax extractor.

9.17

9.16

9.16: Chunks of honeycomb in jars of honey are an attractive way of displaying this product.

9.17: Honeycomb stored in a takeaway food container.

9.18: Preparing the wax. For small quantities of wax capping, melt the capping in a large saucepan bought from an op shop. The pan needs to contain about 3cm of water to stop the wax over-heating and catching fire.

9.18

SAUCEPAN METHOD

You can use a large saucepan from your kitchen if it is no longer being used or find one at your local op shop. Choose a size that will comfortably fit all the wax that you have for melting. Fill this about a quarter full of water and raise the temperature of the water on a stove, slowly adding the cappings until they have all melted. The resulting melted cappings is messy and contains all sorts of unwanted material that will float on top. The beekeeper may need to negotiate the use of the family kitchen for this method and make sure any resultant mess or spilt cappings material is thoroughly cleaned up.

It is critical that the beekeeper remains close to the wax while it's being melted on the stove, to monitor its temperature. If the process is interrupted and the stove left unattended it should be turned off until the beekeeper can return and complete the operation.

Before melting the wax, prepare a large bucket by adding about 5 centimetres of water for the wax to float on while it hardens. Tie some cloth or hessian sacking around the top to act as a filter and set it up outside the house. Do make sure that there is a concave pocket in the cloth or sacking to catch the melted wax so that it will not

9.19: Pour the molten wax into a large pail with hessian sacking tied over the top to filter out the bits of bee and cocoon, commonly called slum-gum.

9.20: Once the wax has solidified in the pail, it can be removed and stored.

overflow onto the ground beside the bucket. Next, very carefully pour the wax into the cloth on top of the bucket. The melted wax will flow through the cloth sieve leaving the slum-gum in the cloth pocket or depression in the material used. Slum-gum is the fibrous black mess left when cappings and comb are melted and filtered; this is particularly so for brood comb as it contains the silk cocoon, defecations and shed skin left behind by pupae as they develop and grow.

When all of the wax has been filtered and the wax both inside the bucket and on top of the cloth has hardened, remove the cloth and hardened slum-gum and either dispose of it in the bin or in the compost. My experience is that although professional beekeepers have the equipment to extract further wax from slum-gum, this is very difficult for a hobby beekeeper and a lot of time can be wasted extracting very little additional wax.

When the wax inside the bucket has hardened it can be removed. If you turn the block of set wax over you will notice some slum-gum has passed through the cloth and has congealed to the underside of the wax block. Cut this off with a hive tool or sharp knife and the remaining wax is usually sufficiently clean to swap for milled foundation from your local supplier.

Since beeswax will be discoloured by pots and containers made of iron, brass, zinc and copper it is a good practice to melt wax only in containers made of aluminium or stainless steel.

If you plan to make candles or furniture polish from the wax, a further level of filtering, or rendering, will be required and the process will need to be repeated again. If the wax is to be used for cosmetic purposes to make face or hand cream, the process will need to be repeated yet again in order to get a higher quality, pure beeswax.

SOLAR BEESWAX EXTRACTORS

Another way to melt the beeswax is to use a solar beeswax melter or extractor. This method has been used successfully over many years and has the advantage that the beekeeper need not stay watching the wax melt but can leave it in the melter/extractor on a warm or hot day and let the sun do its work. A solar wax melter or extractor is usually a square or oblong box painted black or some other dark colour on the outside to attract heat, and white on the inside. Two sheets of glass or clear plastic are fitted to the lid allowing a space between the panes rather like double-glazing. This lid can be opened and shut by the beekeeper when a tray containing frames or wax pieces is inserted or removed.

9.21

Wooden supports are incorporated inside the box to raise the wax-holding tray. The tray should have a lip at one end so that the melted wax can pour into the container provided at the end of the tray. The melter box is placed in a sunny location and tilted at a slight vertical angle towards the sun. The interior of the box acts like a greenhouse and heats up, melting the wax inside, which then runs off and is collected in a container placed inside at the bottom of the tilted box. Check the solar extractor periodically to ensure that it has not run out of wax cappings. Larger solar wax extractors are useful for beekeepers that have space in the garden or have a lot of hives and thus a lot of wax to process. Smaller solar extractors that only hold one or two frames can also be used and are more convenient for the urban beekeeper confined to the average suburban block.

9.22

9.21: An alternative way of processing beeswax is to melt and filter the wax in a large solar wax extractor.

9.22: Smaller solar extractors can also be bought that are more convenient to use and store.

Although Wax Moth would prefer to eat protein sources such as pollen, the moth can partly digest unclean honeycomb. It is important therefore that unfiltered beeswax be processed and cleansed as soon as possible after it has been removed from the frame. Neglect to do this and the moth may completely take over the unfiltered comb leaving a complete mess from which it will be difficult to extract any wax.

COLOUR OF WAX

New honeycomb made by workers is almost white in colour. If there is no brood present, such as in the case of a honey super above a queen excluder, the wax will only slowly yellow with age. The comb yellows in colour due to pollen and also to propolis.

Conversely honeycomb that contains brood quickly goes a chocolate brown colour and over three or four years turns almost black from pupal skins, silk cocoons and brood excrement. Although old dark wax does not look that attractive, it is in fact much less brittle than white wax since it contains a lot more oils. Also, bees appear to prefer the smell of darker foundation and a swarm is less likely to abscond from a hive containing dark foundation than from one containing very light foundation. I have previously mentioned that one of the main management tasks of the beekeeper is to regularly replace drawn frames with old chocolate brown or black comb with clean foundation. A few of the old frames, after replacement with new frames, could be kept to attract future swarms.

Foundation comes in different thicknesses and is usually measured as a number of sheets per kilogram. Foundation typically comes in weights between about 11 sheets/kilogram to 17 sheets/kilogram; 11 sheets/kilogram being the thicker and thus stronger foundation. Which to use will often be a personal choice guided both by price and what is available from your supplier. Thicker foundation is often easier to insert into the frame and due to its extra strength holds together better and is less likely to break when it is rotated in an extractor. Generally, the thicker the wax the better its strength and durability in use. If, however, you propose to cut the foundation out of the frame to extract by hand, use of the thinner product is usually fine as the bee will add its strength when both sides of the sheet are drawn out into comb.

Good foundation should be pleasant to smell since the impurities will have been filtered out of it. As autumn progresses and sales of foundation drop, poorly filtered wax foundation that has been left to stand for some time will start to smell mouldy or earthy. Well-purified wax will continue to have a pleasant smell even after years of storage. Foundation, when left over winter, often forms a white bloom on its surface and many people believe that this is caused by a fungus. In fact, the white bloom occurs due to wax oils that have risen to the surface of the sheet and crystallised. Bloom melts at 39.8° Celsius, much lower than the melting point of beeswax itself, which is about

62° Celsius. Bloom can quickly be removed either by leaving the foundation in the sun for a few minutes or by gently heating the surface of the foundation sheet with a hair dryer. Bloom can also be partly removed by wiping the crystallised oils off the surface with a damp cloth. Winter bloom does not appear to be detrimental to bees or hinder them from drawing out the foundation sheet in the usual way.

If you like eating honeycomb or intend to market this product, it is better to use the thinnest foundation possible since the taste of thick foundation in the centre of a cut comb piece can be unpleasant. The best solution for the beekeeper who wishes to produce honeycomb is not to insert full-size sheets of foundation into frames at all but to put a 1- to 2-centimetre starter strip of foundation at the top of the frame and then leave the bees to draw their own thin, clean, white comb throughout the remainder of the frame. Honeycomb made by bees without foundation is by far the most pleasant to eat.

SUMMARY

- Extracting is an important part of beekeeping and can be performed inexpensively and easily. A simple and inexpensive method is to use a sieve to drain chopped up honeycomb. Beekeepers with more hives may prefer to buy or rent an extractor since these are convenient to use for, say, two or more hives. If you plan to use an extractor, the surface cappings of the honeycomb need to be removed either with a scratcher or with a hot knife. Wax cappings collected as a by-product of extracting can be melted down and sold back to wax foundation manufacturers, or melted down and filtered to make candles, hand cream or face cream.
- Liquid honey is the most popular form of honey although many people also enjoy eating honeycomb.

10.1

10.2

Preparing honey for sale and competitions

PREPARING HONEY FOR SALE

Colonies of bees usually produce much more honey than can be consumed by a family and selling the surplus honey to friends or at markets is a way to repay yourself some of the money you invested starting up as a beekeeper. Most first-year beekeepers do not harvest much honey, often because they became beekeepers after the main honey flow had finished, but also because their new colony of bees was not sufficiently populated with foragers to be very good at collecting nectar. During the second and subsequent years, as the bees learn how important it is to the beekeeper to produce lots of honey, they try harder and they often produce 50 to 80 kilograms per hive, sometimes a lot more in good floral areas during good years!

If you plan to give honey to family and friends the honey can be extracted and bottled in the family kitchen and, apart from general cleanliness, no particular set of hygiene standards will apply. If, however, you plan to sell honey commercially at a market or through a shop, the location where you intend to extract and bottle your honey comes under some very strict state and territory regulations enforced by your local council. These include monitoring and inspections for general hygiene and also a requirement that the working surfaces in the area must be cleanable and not porous. Porous surfaces enable the absorption of water and food, allowing bacteria to grow. While a glass working surface can be used this will not be as durable as surfaces made of stainless steel.

10.1: Preparing comb honey for competitions.

10.2: First cut of honey.

The expense of setting up a suitable separate area to bottle and extract honey is one of the big problems facing hobby beekeepers wishing to sell honey commercially. The best way to find out about food preparation and handling requirements in your state is to attend a food-handling course run by the local council. Courses are usually inexpensive and typically last for one or two days or are held over three evenings during the week. Before anyone is given a licence to operate a honey processing kitchen it is a legal requirement to attend one of these courses. State regulations requiring the considerable expense of providing a separate working area often prevent hobby beekeepers selling honey commercially. If you feel that you have sufficient excess honey and a ready market of buyers it is worth contacting your local council to check what their requirements are since some of these rules are open to interpretation. Another obvious way to avoid the individual expense of setting up a separate area is to form a loose or formal cooperative undertaking with other individual or small producers which would enable the cost of hiring a suitable commercial kitchen to be shared.

LABELLING

Assuming that you comply with local council requirements and have the space to set up a separate extracting and bottling area, your next challenge will be to design and produce your own label. You will want the label to enhance the presentation of your honey and in addition to its visual appeal it must comply with food labelling regulations requiring that certain information is clearly displayed.

The requirements for labelling are complex and as a starting point I would suggest visiting a local supermarket, buying a jar of honey from a reputable supplier and following the type of information disclosed on their label as a guide for your own. The main points to note are that the label must show the major ingredient of a product (100 per cent pure Australian

Nutritional Information

Servings per package: 25
Serving size: 15g

Avg Qty:	Per serve	Per 100g
Energy	212kJ	1416kJ
Protein	0.05g	0.3g
Fat – total	0g	0g
– saturated	0g	0g
– trans	0g	0g
Cholesterol	0mg	0mg
Carbohydrate	12.5g	83.1g
– sugars*	12.4g	82.5g
Sodium	2.3mg	15mg

*Sugars naturally occurring in honey

10.3   Pure Australian Honey

10.3: This label listing nutritional information for honey is typical of one designed and used by a reputable honey producer.

honey) and it must include the contact details for the supplier. The weight of the contents should also be noted. Honey is a complex substance containing many trace elements, however, only the major nutritional components need be listed on the label. If you decide to use the nutritional information displayed on a commercial jar of honey as a guide do be aware that the composition of honey from different floral types varies significantly so the list of ingredients is only a guide for a typical blend of honey.

A small number people that I meet complain that imported or even some Australian bottled honey contains additives that are not listed on the label. The labelling laws for honey, as for all foods, are strict. If the label says 'pure honey' and does not list any further additives then you can be reasonably sure that it contains pure honey. I say reasonably as there have been cases where imported honey has been found to be adulterated with sugar syrup and other fillers. This is not a frequent occurrence and generally a jar labelled 100 per cent pure Australian honey is just that. If a person bottling and selling honey wishes to add a flavour or some other addition — say, extra pollen — this is permissible but the additive must be clearly included on the honey label.

PRESENTATION

Liquid honey purchased from shops and supermarkets is notable because of its clarity and lack of sediment. Supermarkets and other retailers do not want honey to become cloudy or to crystallise on their shelves as the uneducated consumer will interpret these as an indication that the honey is full of impurities or is old, and it will not sell. Consumers equally do not want their honey to become cloudy and start crystallising soon after they have purchased it. To achieve liquid, clear, sediment-free honey, commercial honey producers heat honey to make it more liquid allowing them to pass it through very fine filters to remove any pollen or particles of wax. This also dissolves any microscopic crystals of sugar that larger crystals may form around.

The advantage of home-produced honey is that it is not usually heated, preserving many nutrients and enzymes that are negatively affected by heat. Once consumers taste cold-extracted honey, they realise what they have been missing by opting for the over-filtered commercial product. Some people prefer to buy slightly cloudy honey as it is a sign that it contains pollen, which is believed to be highly nutritious. A well-presented jar of home-produced honey can usually be sold at a higher price than its commercial equivalent.

10.4: Frame honey is commonly available at reputable restaurants during breakfast. Customers can cut off both honey and comb and spread it on toast or mix it with cereal.

PREPARING HONEY FOR COMPETITIONS

Every beekeeper takes pride in both the health of their hives and the quality of the honey produced in those hives. One way to show off the quality of honey is to enter it into one of the many honey competitions that are held every year and to have it judged against honeys from other beekeepers. Although preparing liquid honey for entry into a competition is relatively easy and well within the skill level of even the novice beekeeper, many are reluctant to enter their honey because they are unfamiliar with what is required. This section will therefore cover what is involved in entering honey into a competition.

COMPETITION CLASSES

There are a number of bee products that can be entered into most competitions. The focus will be specifically on the following:
- liquid honey
- granulated and creamed honey
- chunk or cut honeycomb
- frames of honey.

EXTRACTING HONEY AND PREPARATION FOR COMPETITION ENTRY

Liquid honey is the basic product of every hive and is typically the class that has the largest number of entries in a honey competition. Unfortunately, though, liquid honey is the most difficult class to prepare for and, due to the large number of entries, is the most difficult class to win a prize in. Do not be deterred as in both club competitions and even international events such as Apimondia, novice beekeepers have walked away with top prizes against stiff competition from experienced Australian and international beekeepers.

In order to prepare honey samples, the beekeeper could simply fill three or four bottles from the latest extraction. This method, although easy, is not setting the beekeeper up for success in the competition field. Instead, approximately six full frames that have been completely capped or are almost completely capped should be selected. This will ensure that the honey entered will have the maximum density.

When selecting the frames, one way to ensure that all of the honey is from the same floral source or colour is to hold each frame against a bright light and to check that the colour of the honey across each frame, and between frames, is the same colour.

To remove the honey from the frames use either an extractor or a press to squeeze the wax comb until the honey runs out. Pressing the wax combs is preferable since fewer bubbles of air will enter the honey using this method. Once the honey has been extracted, it should be stored for two or three weeks at room temperature, preferably at around 37° Celsius, to allow air bubbles and other particles to rise to the top where they can be skimmed off. See visuals 12.13 and 12.14 in Chapter Twelve on sustainable beekeeping for using a fruit press to extract honey.

STRAINING THE HONEY

Once the honey has settled it is time to filter out any impurities. The liquid honey should be gently poured through a fine filter into another container. Fine nylon filters are available from beekeeping suppliers. Muslin cloth is also good at removing even very small particles and is widely used by beekeepers to filter their honey. To minimise the introduction of new air bubbles into the honey during filtering either of the following two methods are useful.

METHOD 1

Tie the nylon or muslin cloth over the top of the container with a small dip in the top of the cloth like a saucer. Tip the container slightly over onto its side and gently pour the honey into the nylon or muslin. The filtered honey needs to flow gently down the side of the container to minimise the amount of new air bubbles that are introduced.

METHOD 2

Sew the nylon or muslin cloth in a cone shape and tie the cone over the top of the container with the pointed end near the container bottom. Gently pour the honey into the cone, where it will pass through with minimal new air bubbles and filter out most of the residual wax particles.

Whichever method is used, allow the filtered honey to remain in the container for about a week, again at room temperature or preferably at around 37° Celsius, to allow any residual air bubbles or fine particles to rise to the surface where they can again be skimmed off.

A simple and effective technique for removing small particles of wax from the honey is to gently lay a strip of cling wrap across the surface. The wax particles will stick to the cling wrap allowing them to be removed when the plastic is lifted away.

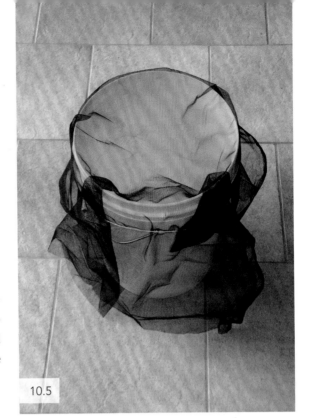

10.5

POURING HONEY INTO A BOTTLE

Containers used in competitions are usually 375 gram or 500 gram glass jars. Each competition will have its own rules so the beekeeper should make enquiries or read the entry information to ensure they know what size and shape jars to use.

For the best chance of winning a competition, good-quality jars should be purchased, not the cheaper variety that often have numerous fine air bubbles trapped in the glass of the jar. Cheaper glass jars often have a greenish tinge to them that will also detract from the visual appearance of your entry.

Before filling the jars with honey, wash each jar in soapy water, rinse well with fresh water and dry immediately with a lint-free towel. This will ensure that the

10.6

10.5: A Nylex net tied over a pail with a hollow in the centre is a convenient way of filtering fine particles from honey. The filtered honey needs to flow gently down the side of the container to minimise the amount of new air bubbles that are introduced.

10.6: A good way to fill jars with clear liquid honey and not get air bubbles in it is to let the honey settle in a pail that has a honey gate at the bottom. The jars can then be filled by letting honey flow through the gate.

jars are sparkling clean and do not have any lint on the inside or outside surfaces that will take marks off your entry. Usually the easiest way to fill a jar is simply to pour or ladle the honey into it. This, however, is not the best way to fill jars for competition entry since the top of the container will contain some air bubbles and other small particles, which will lower the quality of the beekeeper's entry.

A better method is to allow the honey to settle in a container that has a wide valve or honey gate at the bottom. The jars can then be filled by letting honey flow through the gate. The honey at the bottom of the container will be cleaner than the honey at the top since, over a week, most of the air bubbles and other particles will have risen to the surface. Without a honey gate at the bottom, it will be almost impossible to retrieve clean bubble-free honey from the bottom of the container.

If a container with a honey gate is unavailable, an alternative method is to allow the honey to settle in a disposable plastic container, such as an ice-cream container. When the jars are ready to be filled a very sharp knife can be used to cut a small hole in the bottom of the plastic container allowing the honey to flow through the hole and into the jar.

When filling the jars, care should be taken to ensure that they are slightly tilted and that the honey flows gently onto the side of the jar. This will further minimise the number of air bubbles introduced into the honey. Because air bubbles can be introduced into an otherwise excellent honey sample, at every step in the preparation of the final sample the beekeeper will need to constantly and carefully monitor the process, thinking of ways to remove impurities and eliminate new air bubbles. Completion of these tasks is well within the capabilities of even the novice beekeeper if care is taken during the filtering process.

Each jar should be filled to the same height, slightly below the rim of the lid, so that the judge can see the top of the honey without opening the jar. Once the jars have been filled they can be stored in a dark place for a few weeks until the competition. Storing honey in the dark will minimise the risk of the colour of the honey changing before the competition day. When storing the jars, since judges also look at the quality of the jar that the honey is presented in, place some cardboard between the jars to stop them scratching each other.

The majority of competitions require two or three samples of the same honey to be submitted with every entry. If only two jars are required for an entry, three or four samples should be prepared and then, just before the competition, the best two samples should be selected for entry using the judging criteria explained a little later in this chapter. Again, this will maximise the chances of gaining a place or prize in the competition.

If the honey that you plan to exhibit becomes granulated it can be gently heated in a warm water bath or in warm air at a temperature no higher than 43° Celsius. Heating the honey at temperatures higher than this will risk darkening its colour and losing some of its taste.

Each entry class of honey will be restricted to a type of floral source, say, Yellow Gum. If it is necessary to blend honey from different hives to make up the required quantity the beekeeper should make sure that the honey from each hive comes from the same floral source and preferably from the same location.

With a good source of honey, the will to win and the patience to prepare and show honey to its best advantage, by following the steps above a prize is well within the grasp of even the newest beekeeper.

Marking grades given in this book for all classes of hive products are used in the Sydney Royal Easter Show. Check with the competition organiser of the event that you wish to enter what their marking criteria are to ensure the best possibility of gaining high marks with your entry.

LIQUID HONEY

Each competition organiser will publish in advance the judging and marking criteria for their particular competition.

Flavour	25
Density	25
Colour	25
Aroma	10
Clearness	10
Brightness	5
Total Points	100

FLAVOUR

The honey submitted for judging needs to be palatable and free from any tang, harsh or biting taste due to fermentation, acidity or impurities from other floral types. As liquid honey is typically entered as part of a floral type — for example, Yellow Box — the entry needs to be typical of that class since the judge will be familiar with the different floral classes and will also have other honeys from that class to judge it against.

10.7

10.8

10.7: If the honey that you plan to exhibit becomes granulated it can be gently warmed in a warm water bath or in warm air at a temperature no higher than 43°C.

10.8: High marks will be given to the honey with the highest density. This can be estimated by inverting the jar and noting how quickly an air bubble rises to the surface.

10.9

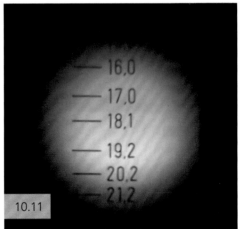

10.10

10.11

10.9: A refractometer works by smearing a small amount of honey across the glass screen; the water content is read out from a scale built into the meter.

10.10: A refractometer being used to measure the water content of honey.

10.11: Reading a refractometer. This honey has a water content of about 18 per cent.

DENSITY

All honeys vary in density and, for the same floral class, this characteristic varies both from location to location and also over time from flowers from the same location. The highest marks will be given to the honey with the highest density and this can be measured either by the use of a refractometer that measures the buoyancy of a glass rod at the surface of the honey, or by inverting the jar and noting how quickly an air bubble rises to the surface.

COLOUR

Many honey competitions have separate classes for dark, medium and light honeys. The colour of honey can be measured with a Pfund honey grader. These graders are relatively inexpensive to buy online and are a good investment for the beekeeper planning to enter honey into many competitions. Dark and medium honey must come within a prescribed colour band, while the light honey class gives top marks to the lightest honey entered in the competition.

WATER CONTENT

Honey should have a water content of less than 20 per cent, otherwise it will ferment and become sour. A refractometer is used to measure water content.

AROMA

Each type of honey has its own characteristic aroma and top marks are given to the entry with the most pleasing aroma, free from fermentation, the smell of being over-heated or any other offensive odour. Each floral class will have its own distinctive aroma — for example, Yellow Gum, Red Gum or Leatherwood from Tasmania — and this bouquet will be taken into account by the judge.

CLEARNESS

Each entry into a competition must be clear of visible impurities and air bubbles and have a sparkling appearance. The cleanliness of the jar will play an important part here and it is critical for the beekeeper to ensure that the jars chosen for use are clean and free from any surface marks such as lint, fingerprints or even air bubbles in the glass of the jar itself.

BRIGHTNESS

Just before the competition slightly heat the jars of honey to increase brightness. Slightly heating the honey will dissolve any sugar crystals and give added brilliance to the sample. Make sure, though, that there is sufficient time for the honey to cool to room temperature before it is judged.

GRANULATED AND CREAMED HONEY

There are often separate classes for these two types of honey in competitions. The main differences between these prepared honeys are that granulated honey has a larger grain size than creamed honey and creamed honey also contains fine bubbles of air.

GRANULATED HONEY

Honey used to make granulated honey should be of the same high quality as used in liquid honey described above. The three major considerations that are used by judges to award points are the smoothness of the granulation, the flavour and the firmness of the set.

Evenness of grain	30
Flavour	30
Firmness	30
Colour	10
Total Points	100

10.12

10.12: Bees on honeycomb.

10.13: Stirring starter into liquid honey to prepare candied honey.

EVENNESS OF GRAIN

Competitions for granulated honey may include both fine- and coarse-grained classes. Fine-grained honey should be smooth with the grain almost indistinguishable to the tongue and with all grains an even size. Fine-grain granulated honey may be made with lucerne or clover. Honey from eucalypts gives a larger grain size.

GRANULATION

In order to prepare granulated honey, a starter is added to initiate granulation, or crystallisation, of the liquid honey. This starter is coarse granulated honey that has either been prepared by the beekeeper or obtained elsewhere. When selecting a starter, it is important for the competition beekeeper to be aware that the consistency of the final granulated honey will mirror the granulation of the starter. For this reason the beekeeper should make sure that the evenness of grain of the starter is satisfactory before using it to prepare competition entry quality honey.

Add about 10 to 15 per cent of starter to the liquid honey that you wish to granulate, mix thoroughly without adding air bubbles to the mix until the entire mix is of the same colour and consistency. Leave this mix in a larger container to settle for 12 to 36 hours at 14° Celsius. Before the granulated honey sets pour it slowly onto the inside surface of the show container making sure that air bubbles do not enter the mix. Leave the jar for about three weeks at 14° Celsius for the granulated honey to set. When the granulated honey has become consistently firm throughout the container it is ready for competition entry.

FLAVOUR

The flavour of the granulated honey will represent the floral source that the liquid honey comes from.

FIRMNESS

The granulated honey mix needs to be firm but easy to spread. The starter used will strongly influence the firmness of the beekeeper's honey entry so this characteristic should be considered when selecting a suitable starter.

COLOUR

The colour should be between cream and off-white, but will be determined by the colour of the liquid honey used.

CREAMED HONEY

The process used to prepare coarse granulated honey also applies to smooth creamed honey. The main differences are that the starter needs to be fine grained and that the colour of the final product needs to be white. The scoring scheme for both granulated and creamed honey is usually the same.

When making creamed honey the starter should be added to the liquid honey in such a way as to incorporate air into the mixture, as this will give the creamed honey much of its white colour. Since creamed honey is very viscous, a household grade mixer is likely to overheat and burn out if used. If available, it is better to use a dough mixer and to beat the honey until the mix will not go any whiter.

CHUNK HONEY

Chunk honey is a cut-out section of capped honeycomb that is placed in a jar of liquid honey.

Appearance	25
Colour	25
Flavour	20
Density	20
Clearness	10
Total Points	100

The beekeeper should select a frame containing capped honey with the capping smooth, white and pleasing to the eye. The wire should be removed from the frame, as this will make it easier to cut out a nice section of comb. Use a hot knife and cut out the selected section of comb, placing it gently onto a wire grille so that the honey at the edges, together with any pieces of wax, drains off.

Next the chunk is placed into a jar ensuring that no force is used to pass the chunk through the opening. With the jar slightly tilted, high-quality liquid honey, prepared as previously described above, is gently poured onto the side of the jar making sure that no bubbles enter the honey. The liquid honey and comb need to be of the same floral type. The comb should rest in the honey with the top of the chunk resting just below the lid.

FRAMES OF HONEY

As well as chunk honey, there are often classes for frames of honey. A frame should be selected that is brood and pollen free and with surface capping of the honey that is white, smooth or regular, and pleasing to the eye. Since the frame that is used to store the honeycomb is also included in the competition make sure that the frames selected are well assembled and are of good quality.

10.14

10.14: To make a good presentation of cut comb honey, let the cut comb drain on a wire grille so that all of the honey is removed from the sides of the comb. Use tweezers to remove particles of wax comb for a smooth edge. This will make the display much more attractive to the judges.

Fullness	30
Colour of cappings	30
Evenness	20
General appearance	20
Total Points	100

The quality of a frame of honey is very much dependent on the hive that it came from. When selecting hives from which to draw frames of honey for competition, look for colonies that store honey with a small amount of air under the cappings as this will give a whiter appearance than those frames that do not have air under the cappings. Once the beekeeper determines a frame of honey is of competition quality it should be removed immediately capping has finished so the bees do not get time to walk over it and discolour the surface. For this reason when selecting competition frames the beekeeper should keep away from colonies that produce a lot of propolis as this may discolour the wax and reduce its visual appeal. As part of the selection process the frame should be held up with a strong light behind it and a check made for impurities in the honey or wax such as pollen, propolis or brood as these will severely reduce the chances of winning a prize.

The judge will also look at the drawn frame. When selecting frames of honey make sure that the wooden frame is clean and does not contain propolis or other surface markings.

10.15

10.15 Frame imperfections can lose points from an otherwise prize-winning frame of comb honey.

FULLNESS

Judges give the highest marks to frames that have been well drawn out.

COLOUR OF CAPPING

Cappings should be white with no differences in colour across the surface. Differences in colour may indicate different floral types stored across the frame and this will lose the competitor valuable marks.

EVENNESS

The surface of the wax should be even with no depressions or rises.

GENERAL APPEARANCE

It is important that both the capped honey and the frame are pleasing to look at. When setting up a super to prepare quality entries for competition the beekeeper should ensure that only new frames are used, as this will give an overall better appearance to the entry being judged. To ensure this, the beekeeper interested in entering competitions should inspect hives early in the season and any defective or dirty frames should be replaced with new frames. If the frame selected for competition entry has some minor blemishes these can be removed very carefully with fine sandpaper.

In order to protect the competition frame against damage, a frame holder may easily be made or purchased from a beekeeping supply store.

SUMMARY

- Honey prepared at home tastes far superior to honey purchased at a supermarket. If you have a surplus of honey it is usually straightforward to sell this to family, friends or co-workers. To do this there are no stringent legal or health rules apart from the common sense requirement of general cleanliness. If however you plan to sell honey commercially you will need to obtain certification for your honey extracting and bottling area from your local council.
- Many beekeepers would like to enter their hive products in a competition and this is a lot easier than many novices believe. With a small amount of care and careful monitoring of the exhibit preparation process, even amateur beekeepers have succeeded in winning major prizes against stiff national and international competition.

11.1

Rearing queens

There is no doubt that the success of the honey harvest each season depends on both the quality of the queen and the available flora. Without high-quality workers produced by a viable well-mated queen the overall health of the colony and resultant honey harvest will not be the best that you can get. Although the desire to harvest the maximum amount of honey is one of the main reasons for rearing and introducing a new queen, there are many others. These include:

- learning new skills as a beekeeper
- the satisfaction of rearing and introducing your own queens
- selecting strains of queens that are better suited to the conditions in your area
- to make money selling queens
- the added prestige of being able to successfully raise queens rather than just keeping bees
- to overcome the common difficulty of not being able to buy a new queen when you need one
- to learn more about the science and genetics of bees
- the pleasure in observing a fascinating natural process taking place within your own hives
- engaging in a shared activity with other beekeepers, and
- just for the heck of it.

11.2

11.1: Workers covering the cage of a recently introduced queen.

11.2: Marking a queen.

There are many well-proven methods used to rear queens, each with its own advantages and disadvantages. Some methods suit the hobby beekeeper with only a few hives while others are better suited to the serious or professional beekeeper with many hives and who needs to rear large numbers of queens each year. Queens can be reared almost anywhere in Australia, from Tasmania to Queensland and across to Western Australia. Although there are many commercial queen rearers in Queensland and New South Wales, there are also very successful and reputable rearers in the southerly states of Victoria and Tasmania.

In this chapter I will explain the hive splitting method for rearing queens. This is easy to execute and the technique has the added advantage of being a useful tool that can also be used to manage swarming.

If you plan to rear queens you may select any suitable hive, split the colony, and wait and see what the characteristics of the resulting queens are. Although this wait and see approach is easy to use, it will not provide you with the high quality queens that a more deliberate approach will allow.

Assuming that you have several colonies that can be used to rear queens, select the colony that over the past season has displayed favourable characteristics that you would like to foster in your next generation colonies. These may include:

- superior honey production
- disease resistance
- minimal aggression
- good overwintering with strong build-up of brood in the spring, and
- a reduced tendency to swarm.

By selecting a colony that manifests any of these characteristics you should be successful in producing high-quality queens. With care and practice, refining your queen-rearing skills over a number of seasons will ensure that the colonies in your apiary are headed by quality queens that possess the optimum characteristics for your area and needs.

Before discussing queen rearing in more detail, however, it is necessary to recap the reasons why a colony make a new queen under natural conditions and what is the optimum situation for this to be successful.

THE THREE REASONS FOR A COLONY TO PRODUCE A NEW QUEEN

The three reasons a colony may decide or need to raise a new queen are discussed in Chapter One on the life history of the honey bee. A sound understanding of these reasons is needed if you are to successfully rear queens. Without repeating the detail given earlier, the three reasons that a new queen may be raised by a colony are swarming, supersedure and emergency conditions.

In addition to what is going on inside the hive the second most important consideration in rearing queens is the nectar and pollen resources available to the colony. Rearing queens in any sort of number requires the production of large amounts of royal jelly and this in turn requires large numbers of young bees that have themselves been reared on a high-quality diet consisting of a variety of pollens and an adequate nectar flow.

In the event that floral conditions are not sufficient at the time that you wish to rear queens it is possible to supplement those conditions to a degree. Both nectar and pollen supplies can be augmented with substitutes from time to time. White sugar mixed in a 1:1 ratio with water to make sugar syrup provides a very close approximation to nectar and has an identical effect on brood rearing by the colony. There are many commercial irradiated pollen supplements available and these typically contain a blend of soy flour, yeasts and small quantities of amino acids and minerals. The other source of pollen is bee collected pollen that has been irradiated with gamma irradiation or collected from your own hives that you know are free of disease.

When rearing queens the following steps are important. Firstly, the hives used to produce the queen cells must be strong, often much stronger than needed for collecting honey. This may entail merging two or more colonies, each of which would have been perfectly adequate if collecting honey, into a single very strong colony. During the process the size of the hive is also reduced to one or two boxes to ensure a crowded colony.

When raising crowded colonies like this during the spring for queen rearing, great care must be taken to prevent swarming or all of your efforts will have been in vain.

Regular inspection of the brood box is essential to identify and remove any swarm cells before they are capped.

Once you have decided which colony in your apiary possesses the optimum characteristics that you would like to propagate using new queens, you will need to manage that colony so that it is very strong (contains many worker bees), and is well fed, resulting in good larvae production and nursery bees producing a plentiful supply of royal jelly.

HIVE SPLITTING

Hive splitting is the easiest of all methods for raising queens. The steps involved are easy to follow and can quickly be mastered by even an inexperienced beekeeper. The established queen in the brood box must be young and vigorous to minimise the risk of swarming. A young vigorous queen will also lay many more eggs to keep the colony strong while the introduced brood cells are developing in the top brood box.

STEP 1

Take a strong two-brood box hive with young brood in both boxes and physically separate the two brood boxes into two separate hives. Assume that the queen is in the bottom brood box. I usually leave the bottom brood box in the same location. I will refer to this brood box as hive 1. Shake most

11.3: The hive split method for rearing queens involves taking a strong, two-brood box hive and making two hives out of it. Leave the old queen in Hive 1 and ensure that there are plenty of very young larvae in Hive 2. The workers in Hive 2 should start making new queen cells overnight.

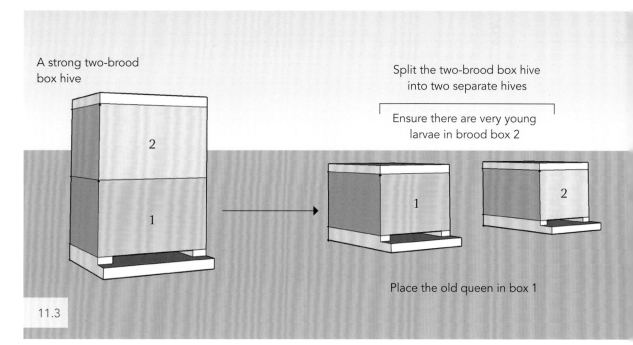

A strong two-brood box hive

Split the two-brood box hive into two separate hives

Ensure there are very young larvae in brood box 2

Place the old queen in box 1

11.3

11.4

11.5

11.4: Preparing a two-box hive to rear queens.

11.5: Instead of splitting a hive, this beekeeper is placing some frames of young brood, capped honey and pollen into a nucleus. A plentiful supply of workers was included in the nucleus to rear new queens and to care for the existing brood.

11.6: Workers raising new queens in a queenless hive.

11.7: A nursery bee feeding royal jelly to a queen larva. A nursery bee entered the cell about every 30 seconds, stayed 2 or 3 seconds, then left.

11.6

11.7

of the bees from hive 1 into the second brood box, which I shall call hive 2. Hive 2 can either be placed next to hive 1 or much further away. When you shake most of the bees from hive 1 into hive 2, all of the foragers will quickly return to the original location, hive 1, balancing the number of bees in both boxes.

When preparing the two hives place the greater amount of sealed brood into hive 2; this will compensate for the drift of foraging bees back to the original hive entrance. The old queen remains in hive 1.

Note that it does not matter which hive the queen is in. If you know which box the queen is in place at least one frame with eggs and less than one-day-old larvae in the box that does not contain the queen. If, on the other hand, you do not know which box the queen is in ensure that both boxes contain eggs and less than one-day-old larvae. If you have not previously found the queen it does not matter, as the workers will raise the new queens in whichever brood box is queenless.

STEP 2

Wait three to five days and inspect whichever hive is queenless — or both hives if you do not know which of the two brood boxes contains the queen or, in other words, is 'queen-right'. The brood box that is queen-right should have new eggs and larvae while the brood box that does not have a queen will not contain any eggs or very young larvae and should show evidence of queen cells being made by the workers.

STEP 3

The next step is to prepare a mating nucleus to take the ripe queen cells. A ripe queen cell is a capped cell out of which a queen will emerge in less than 24 hours. Each queen cell left in the split brood box will need its own mating nucleus and these need to be prepared in advance. Two or three 4- or 5-frame mating nucleus can be prepared from one strong eight- or ten-frame hive. Into each of the mating nucleus place two frames of brood, one frame of capped honey and one frame of pollen. Any remaining space can be filled with an empty frame of drawn comb. The old queen from the eight- or ten-frame hive can be placed into her own nucleus for future use.

STEP 4

Ten days after the two brood boxes have been separated, about two days before the first queen emerges, it is time to move each queen cell to its own mating nucleus. Timing is important in removing each queen cell and placing them in their own nucleus to develop into mature queens. If you leave the capped queen cells together in the hive for too long the first queen to emerge will kill the other developing queens in their cells before they can be moved to their own nucleus.

11.8

11.9

11.10

11.11

11.12

11.13

Using a sharp modelling knife, cut out each capped queen cell, being very careful not to expose the resident queen, and then place it into its own mating nucleus. To do this, remove each capped queen cell from the brood box and press into the comb at the top of a frame of brood in the nucleus. Each queen cell removed needs to have its own mating nucleus if the maximum number of queens is to be reared during a hive split. Queen cells are fragile and vulnerable to extremes of temperature and drying out. Mating nucleus should ideally be close by and the transfer of cells should not be done in hot and dry conditions. Close the nucleus and leave it for a week. By this time the queens should have emerged from the cells although they may not have had the time to develop sufficiently to go on their mating flight.

11.8: Cutting a ripe queen cell from a frame to transfer it to a mating nucleus.

11.9: Attaching a ripe queen cell to a frame from the mating nucleus.

11.10: Preparing mating nucleus for the queens to hatch in before going on their mating flights.

11.11: A queen emerging from her cell unaided by workers.

11.12: Once the queen has emerged from her cell, she wanders around unnoticed by other bees. It will take a few days before the new queen starts to release queen pheromone and the workers take any notice of her.

11.13: Mating nucleus used by a queen rearer in northern Victoria.

11.14: Queen cages are used to hold a new queen and her worker bees when they are to be placed in a new colony.

11.15: A queen cage being used to insert a new queen into a queenless colony. Note that the entrance to the cage is facing upwards. This is to stop any dead bees inside the cage blocking the exit if it were to be facing downwards.

NEW QUEENS EMERGING FROM THE NUCLEUS HIVES

About one week after emerging as virgin queens from their capped cells the new queens will go on their mating flight.

A few days after the mated queen has returned to the mating nucleus she will start laying eggs. You can leave the laying queen in the mating nucleus for several weeks while the colony builds up in numbers of workers. Once the mating nucleus is full of brood, honey and pollen, you can transfer the colony to its own eight-frame or ten-frame brood box and you now have a new hive to add to your apiary.

Alternatively, if the aim of rearing queens was to re-queen existing colonies that need new, young, vigorous queens, find and remove the old queen before introducing the new queen using a queen cage. The entrance to the queen cage needs to be blocked with candy. It will take two to three days for the candy to be eaten allowing the queen to escape and take up her role in the colony. If the queen is introduced immediately into the colony the resident workers will not recognise her smell and will kill her. In the two or three days that it takes for the new queen to escape her cage, the resident workers will have become used to her queen pheromone and will usually allow her to become part of the colony without harming her.

Make sure that the queen cage has very small gaps in it or the resident workers from the colony will enter the cage and kill the new queen. Alternatively, even if the gaps in the walls of the queen cage are too small for workers to enter, they may be sufficiently large for workers to bite and chew the new queen's feet through the gaps and thus kill her before she can leave the cage. Infrequently there is also the situation where workers will group and surround a cage, 'balling it' and smothering the queen inside.

So far I have only discussed the importance of selecting quality colonies from which to rear queens. Equally important, but much more difficult to manage, is ensuring that the virgin queen on her mating flight mates with quality drones from other colonies.

Professional queen rearers often manage this by flooding the area around their queen-rearing apiary with colonies containing large numbers of quality drones. Although this practice is achievable by professional queen rearers it is not achievable by hobby beekeepers or small-scale queen rearers. Hobby beekeepers who rear queens can only hope that the virgin queen on her mating flight will mate with a sufficiently large number of drones to make the genetic diversity of her offspring strong, producing healthy, viable colonies.

QUEEN CANDY

If you plan to send queens by mail to other beekeepers you will need to buy queen cages as well as make queen candy to block the entrance to the cage.

Queen candy is different to the sugar candy used to feed colonies over winter in that queen candy is a fairly soft, putty-like mixture of honey and icing sugar while sugar candy is a solid block of white sugar. To make queen candy take a small amount of irradiated (disease-free) liquid honey and mix into it some finely granulated icing sugar. Keep adding and kneading in icing sugar until the resulting mixture has the consistency of putty. There are several types of queen cage available from suppliers and depending on the design you purchase the queen can be introduced in one of the following ways. For a two-entrance hole cage, begin by blocking one of the entrances to the queen cage with the queen candy. Next, place inside the cage the newly mated queen with about six worker bees to look after her until she is able to leave the cage. Once the queen and her escorts are inside the cage, the second entrance can usually be closed by a plastic plug. If your cage is of the type that has one entrance you will need to introduce the queen and her escorts before carefully inserting the candy plug.

SUMMARY

- Rearing queens is a valuable skill to learn as it will enhance your overall beekeeping knowledge and expertise. There are several methods of rearing queens and the easiest to master is the hive splitting method in which a single strong hive with brood in two boxes is separated into two independent hives. Properly managed, several new queens should emerge in one of the hives about fourteen days after the split.
- High-quality queens can be reared by selecting a queen from your apiary that has, during the previous season, either produced the colony that stored the most honey, was the least aggressive, or was more disease resistant.

12.

12.1

Sustainable beekeeping

There are a growing number of beekeepers who practice what is loosely termed 'sustainable or natural beekeeping'. The philosophy of sustainable beekeeping differs from traditional hobby and commercial beekeeping in that practitioners devote a lot of thought and energy into providing what they believe are the optimum conditions to ensure their bees' existence is as close to that of a naturally occurring colony and as stress-free as possible. There is a view that bees 'know what they are doing' and should be left, with minimal interference, to get on with their business in the hive. The view is of a cooperative partnership between the beekeeper and the bees, with honey production being a secondary consideration.

Beekeeping is seen ideally as a hobby or a small cottage industry and the commercial practice of many hives of bees foraging in a single area is seen as undesirable, both from the point of view of the bees and also the excessive competition created to many native pollinators in the area. Left to their own resources smaller numbers of bees would inhabit the same area and the number of hives would be regulated by the availability of food and water supplies. Many natural beekeepers believe that a large number of the pests, diseases and compromised quality of life of today's bees, including the well-publicised Colony Collapse Disorder (CCD) are largely due to, or certainly greatly contributed to, by current beekeeping practices. See p. 296 for more information about Colony Collapse Disorder.

There are various ways in which modern apiculture can cause stress and resultant poor health in bees. One of the more obvious practices causing nutritional stress is monoculture, where large areas of land are devoted to the production of a single crop. Notable examples are canola (rape and canola are the same thing) or almonds. Bees need a variety of pollens

12.1: The extensive use of monocultures in agriculture has become a major problem for all species of bees. This photograph of a monoculture in Canada shows the difficulty bees have in getting a varied diet. Once the canola has stopped flowering there may not be any available food source for local bees.

and nectars to provide the range of minerals and amino acids required to develop and raise healthy brood and to achieve innate and physical vigour as adults to perform the many duties demanded by the colony. Large areas planted with a single crop often do not allow bees to collect the variety of foods needed to sustain optimal health. Periods of monocultural exposure often supplemented by artificial feeding produce colonies more susceptible to disease, and both physical and nervous system distress.

A fact sheet from the US Department of Agriculture on surveyed bee losses during the winter of 2012–13 records colony losses of around 30.5 per cent. A difference noted by the research scientist conducting the survey was the increase in colonies dwindling away rather than suffering classic CCD characteristics. Of interest is that beekeepers who took their bees to California for the almond pollination reported higher losses than beekeepers who did not take part in the pollination. Nearly 20 per cent of those beekeepers who moved their bees to pollinate Californian almonds lost 50 per cent or more of their colonies. The beekeepers participating in the survey managed around 600,000 colonies representing nearly 22 per cent of the US colony count of around 2.62 million.

While poor nutrient uptake through an inadequate diet causes stress to bees, in many regions this is outside the control of the beekeeper. This is particularly so for those rural beekeepers who rely on the often fickle flowering of native trees and shrubs which are very dependent on climatic conditions. For the urban beekeeper, whose bees are able to forage from a wide range of native and exotic plantings throughout the local community, bee nutrition is much easier to manage. For the hobbyist with a few hives much can be done in the home garden to make it attractive and productive as a source of food for bees. Many beekeepers will also make an effort to locate hives near

12.2

12.2: Urban gardens often have a wide range of exotic and native flowering plants that can provide bees with the range of pollens that they require. These gardens are often well watered and are not as subject to drought as country areas.

a wide range of flowering plants even if this is inconvenient due to distance and difficulty of site access.

A further major cause of stress to the bee is migratory beekeeping as practised in Australia and the United States. Commercial beekeepers, and some hobby beekeepers, move their hives every few weeks to a new location to follow the blooming pattern of local flora. As part of the practice of migratory beekeeping, hives can be loaded onto the back of a truck and shipped up to several hundred kilometres either overnight or during the day. Apart from the stress of being moved and the need to reorient themselves to a new location every few weeks, colonies are additionally stressed by being either confined to hives during their move or, in instances where hives are moved with the entrance open, some foraging bees will escape and not return, thus reducing the number of critical food gatherers in the hive.

In nature, if a bee colony takes up residence in, for example, a tree cavity, in the absence of human intervention or fire, nests may remain undisturbed for a number of years. Many of these feral colonies remain at the same location; they swarm regularly but will eventually abandon the nest, perhaps due to lack of space. If the colony is in a tree hollow, abandoning the nest may be caused by the tree decaying and becoming more open to the elements, ant invasion or a lack of food in the area. There are always exceptions and some very old trees with large hollows have supported viable colonies for a decade or more. If a suitable nest site is abandoned by its colony it is not unusual for the site to be re-occupied by later swarms.

Modern beekeeping practices since the invention and patent of the Langstroth hive, while a vast improvement on skeps and clay pots in which colonies were regularly destroyed in order to obtain honey, do nevertheless pose challenges to the wellbeing of the colony housed within it.

With conventional beekeeping, hobby beekeepers, particularly novice beekeepers, are encouraged to open their hives frequently so that they can sight the usual brood pattern, find the queen and generally get used to viewing a colony in good health so that later on there is some basis of comparison if the hive is opened and there is a change indicating a possible problem. During the swarming season again the beekeeper is encouraged every week to check for queen cells and, if present, to either destroy these or artificially swarm the hive by splitting it into two colonies. Natural beekeepers view over-frequent inspections as unnecessary and disruptive to bees, and many believe that much can be ascertained about the wellbeing or otherwise of a hive by careful repeated observation of the returning bees and the general outward appearance of colony behaviour outside the hive.

While natural beekeepers regard the frequent manipulation of moveable frames as an unnecessary disruption to the life of bees, across Australia, Europe and North America there is a legal requirement for removable frames to be used so that apiary inspectors and beekeepers can easily check for the presence of disease. Indeed, the inability of earlier generations of beekeepers who used skeps, clay pots or other traditional hives to closely inspect their colonies was a major impediment to checking the health of their bees. In practical terms there is a trend among natural beekeepers to use top bar hives to house bees. To minimise disruption of the colony during inspections, some top bar hives have transparent acrylic sheets built into the walls. This allows the beekeeper to carry out visual inspections of the inside of the hive without disturbing the bees. The acrylic window or sheet will have a thick wooden door that can be closed between inspections so that the bees will not be disturbed by light continually entering the hive.

This is a major point of difference between traditional beekeepers and natural beekeepers who are strongly of the opinion that the hive should remain undisturbed as much as possible. There are a few reasons for this: firstly, the smoking and opening of a hive to remove frames is a significant disruption to the colony and the bees may take days to get over the intrusion before returning to their normal pattern of activities. Secondly, by opening the hive the beekeeper is altering both the nest scent and ambient temperature of the colony and this will take time and energy by the bees to return to normal. Also, brood exposed to extremes of heat or cold may either die or be subject to developmental problems during their growth and will not function properly as adult bees.

Honey bees draw a variety of cell sizes when they make honeycomb, ranging from smaller cells at the top of the comb to lay worker eggs and to store honey, to larger cells at the bottom of the comb to raise drones. Natural beekeepers believe that the use of foundation embossed with a single cell size inhibits the bees from determining the optimum layout of the comb and hive, leading again to stress within the colony. By using milled foundation with smaller cell sizes suitable for worker bees and not drone bees, traditional apiarists ensure that there are a higher proportion of workers in the hive than would be the case in a feral or naturally occurring colony. Natural beekeepers argue that it is better to leave the selection of cell size to the colony. In this way the workers themselves will decide the optimum cell size and this size will vary across different parts of the nest and during different times of the year. Allowing the bees to make their own comb has a cost in reduced honey production (approximately 1 kilogram of wax takes 8 kilograms of honey to produce). But the reward comes in the form of healthier, more contented bees using clean, fresh comb, free from accumulated impurities and, of course, disease.

As previously mentioned, migratory beekeeping is practised by professional beekeepers in Australia and North America and although this practice is seen as maximising the amount of honey produced by each hive it causes significant stress to the bees. Moving many hives close together, particularly hives brought together from various parts of the country for pollination purposes, is a very effective way to transmit disease and pests widely and very quickly. Small Hive Beetle is a pest whose incursions have been aided by grouping both unaffected and affected hives in close proximity.

To use the pollination example again, when two colonies of bees with the *Varroa* mite were detected at Port of Newcastle in June 2022, the New South Wales's Department of Primary Industry and local beekeepers mounted a strong effort to trace and kill all colonies in the vicinity. If the same *Varroa*-infested colony had been first detected in north-east Victoria during almond pollination in August, where thousands of hives have been trucked from across eastern Australia, there might have been very little point in attempting to do anything about the mites. Any hives removed from the pollination site prior to detection would have soon been distributed widely and the mite with them.

Around one hundred years ago, diseases of honey bees were localised and bee populations globally were vulnerable to perhaps one or two diseases. Bees developed good resistance to these diseases which had been present in their particular locality in

Europe for hundreds of years or even longer. Migratory beekeeping practices, while seen as an economic necessity for many beekeepers in the short term, may effectively have contributed to the creation of many long-term problems which will cause great economic loss, put strain on the industry and also severely compromise the bee it depends upon.

Since the mid-twentieth century, when *Apis mellifera* was transported across the globe on an unprecedented scale, bee diseases have spread exponentially into local bee populations that have little or no in-built resistance to them. The *Varroa* mite was initially a minor pest of the Asian honey bee, *Apis cerana*. The two lived together for hundreds of years and *cerana* developed a tolerance for the mite and the ability to keep it under control using effective grooming. When *Apis mellifera* was introduced into Korea and Japan, *mellifera* came into contact with *cerana* bees carrying *Varroa* mites and the mite soon jumped species from *cerana* to *mellifera*. Commercial *mellifera* beekeepers in Asia are believed to have used the Trans-Siberian Railway to ship infested *mellifera* hives back to eastern Russia and from there into Europe, resulting in the disastrous situation we have today where *Varroa* is considered the biggest problem of honey bees in the world and by far the largest cause of losses to beekeepers.

NATURAL BEEKEEPING HIVES

Natural beekeepers are not always fans of the Langstroth hive, although many commence their beekeeping life using them. Variants of the top bar hive are usually preferred for a more natural approach to beekeeping. The top bar hives usually favoured can be divided into two main designs: Warré vertical top bar hives and Kenyan and Tanzanian horizontal top bar hives.

The design of both of these types of hive will be explained shortly but essentially their use and appropriate management are believed by their adherents to minimise disruption to bees and to provide a more natural nest environment than the Langstroth hives.

There is no doubt that a thicker wooden wall for the hive, as recommended by Abbé Émile Warré (see p. 194), provides better insulation for the colony and protects the bees from external disturbances. Langstroth hives purchased in Australia are generally made of wood that is 22 millimetres thick; this is a tradition from times when 7/8-inch thick wood was readily available from saw mills and hardware shops. To

12.3

12.3: A skep hive next to a Warré hive. Skep hives, although the traditional symbol of beekeeping, are illegal to use in Australia since frames of brood cannot be removed to check for disease.

12.4: The author's brightly painted Warré hive on the right.

12.5: A Kenyan top bar hive.

12.6: A Warré hive.

obtain 22-millimetre-thick wood today to make your own hive is difficult since few mills cut to this size and you will need to buy large quantities of this wood to keep the cost down. Since this is beyond the resources or needs of most hobby beekeepers the next best option is to buy wood from the local hardware store, which is usually 19 millimetres thick.

WARRÉ HIVES

Abbé Émile Warré (1867–1951), the French inventor of the Warré hive, believed that the optimum cross sectional area of a hive should be 300 millimetres x 300 millimetres square, 200 millimetres deep and that the hive should be made out wood preferably up to 35 millimetres thick for better insulation against heat and cold.

Of course many beekeepers are of the opinion that their way of keeping bees is the only way and that all other methods are harmful and ineffective. Unless something has been lost in translation this attitude is apparent throughout Warré's book on the construction and management of the Warré hive, often called the People's Hive.

Warré has little tolerance for even the smallest change to the design of his hives or his management methods for keeping bees in them. In fact, bees of their own accord will make nests in a range of natural and artificial environments from hollow tree trunks, wall cavities, roof spaces, barbecues, and the odd possum or bird box. To state categorically that a bee prefers a box exactly 300 millimetres square and 200 millimetres deep is, to me, too dogmatic. In fact, many modern proponents of the Warré hive have devised their own, or adopted others adaptations to the original plans of construction but generally adhere to the management methods explained by Warré. To be fair to Warré, when he designed his hive he took into account what, in his view, was the optimum size of hive that the bees could manage taking into consideration the long cold winters experienced in Europe.

A difference between the original design for the Warré hive and today's version is the removable frame. In the 1940s, when Warré specified the design, frames were fixed and the honey could only be extracted by breaking the comb. Warré was of the opinion that the modern removable frame was not a help in the early detection of disease but a major contributing cause. His view was that all that is needed to monitor the health of a colony is to closely watch the comings and goings of its resident bees to and from the hive entrance. Today, to comply with regulations for hive inspection for disease requiring moveable frames, a modern Warré hive frame includes removable top bars but not a bottom bar.

Warré believed that using a bottom bar on frames restricted the ability of the bees to make natural honeycomb according to their needs. A thin starter strip of foundation is used underneath the top bar of the frame since the bees know intuitively the best shape and cell size for their colony. Interestingly, there have been recent variations on Warré's original design with operators of these hives incorporating side bars on frames similar to a Langstroth hive. The side bars prevent the comb being attached to the side walls of the hive box where it may be severely damaged and break off when the frames are removed. In countries where the *Varroa* mite is a problem there have also been modifications incorporated into the Warré hive base.

The design of the Warré hive also differs from a Langstroth hive in that instead of the top super box touching the lid, in the Warré hive a smaller buffering box is placed between the lid and first super box that provides additional insulation for the colony. This box is about 100 millimetres deep and is called a quilt as this term clearly explains its use. The quilt box contains a floor made of hessian cloth or aluminium flyscreen wire. The box is filled with dry leaves, straw, wood shavings or crumpled newspaper to provide insulation and the hessian or mesh wire attached to the bottom of the quilt box prevents the insulating material falling into the hive. The hessian or aluminium mesh floor very effectively allows damp air to circulate into the dry insulating material in the

12.7: A natural beekeeper holding a frame from his Warré hive.

12.8: The quilt box on my Warré hive. My design is slightly different from that specified by Father Warré in that the quilt box is filled with crumpled newspaper or dry gum leaves. Flywire stops the leaves from falling into the hive.

quilt box where it is absorbed, keeping the colony free from condensation. Since a lot of dampness can be absorbed by the insulating material, the beekeeper needs to check that this is not becoming mouldy and will need to periodically replace it with fresh dry insulation as required.

Another major innovation of the hive is the design of the lid. Warré believed that the traditional flat metal lid was too disturbing for the bees since rain falling on it causes a drum-like noise inside the hive. Warré proposed two designs for lids, one of which is probably beyond those beekeepers without access to woodworking tools and facilities. The second, a much simpler design, can easily be made by most beekeepers with a little time and some basic woodworking tools. For all practical purposes the simpler design of lid for the Warré hive is effective and there is little to be gained by using the more sophisticated version of the lid apart from the inclination of some beekeepers to experiment with hives using both.

A simpler lid again which would be effective in Australian conditions is to use the same flat roof design that is found in traditional Langstroth hive but to replace the metal cap with a thick wooden top that will not transmit the sound of rain so easily. The lid in this situation should be designed to be telescopic and not the migratory type of hive lid, as telescopic lids provide much better insulation and protection from the weather than do migratory hive lids. Wood to replace the metal cap can easily be sourced from scrap wood or from local building suppliers. If there is any difficulty sealing the joins of the lid the beekeeper can cover the lid with tar paper obtained again from a builder's supply outlet.

As with the Langstroth hive, a mat is placed on the top frames of a Warré hive, however the material used is hessian cloth soaked in flour and water and then ironed flat when dry.

The design of the base, apart from being nearly square, is the same as for the Langstroth hive. To ensure later ease of handling of the Warré hive lifting cleats need to be attached to the sides of each box.

Management of the Warré hive is quite different to that of a Langstroth hive. When a new box is added to the Warré hive it is not placed on top of the existing hive boxes but is instead inserted at the bottom of the hive, a process more traditional beekeepers would call under-supering. This practice is referred to as nadiring in Warré terminology.

In the Warré hive set-up there is no queen excluder used and the queen is free to move throughout the hive. In their natural state, bees will build comb downwards as required, so it makes sense to provide additional space for the colony to expand at the bottom of the hive. This immediately raises the interesting question of how to lift a number of full boxes to slide a new empty box onto the hive base. Many books on Warré beekeeping show a rather sophisticated lifting system that seems on the face of it, or to me at least, to be totally impractical since it is beyond the capability of most beekeepers to either make or to obtain one of these lifts. The hobbyist with any carpentry skills or a bent for invention will relish facing and sorting out this problem, however, as it has not proved too much of an obstacle for an increasing number of beekeepers choosing this type of hive. Various websites provide many ideas and designs for Warré hive lifts and the interested reader should do a Google search for suitable plans.

Many natural beekeepers who own a Warré hive perform under-supering, or

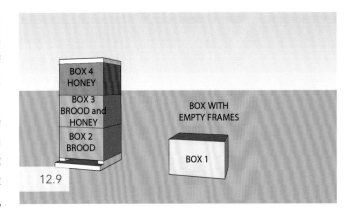

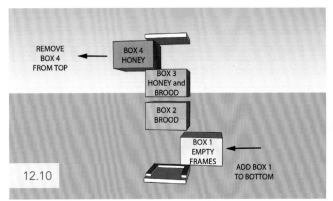

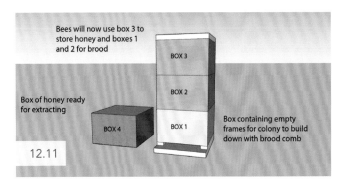

12.9: Part of the philosophy behind the Warré hive system is only to remove the top box since this contains honey. At the same time a new box is added at the bottom containing only frames. This process is called *nadiring*.

12.10: The top box is removed and a new bottom box is added.

12.11: The nadiring process has been completed.

12.12/12.13: Using a fruit press to extract honey from a frame. This is a traditional method of extracting honey from honeycomb. Using a fruit press, a lot of the coarser wax particles are retained inside the press.

nadiring, by removing all of the boxes off the base, adding an empty box with frames, and then returning the other boxes and placing them on top of the new bottom box. While this may be the only way to achieve the addition of a bottom box in the absence of a lifting device, it is in itself very disruptive to the colony of bees and goes against the principle of minimal disturbance.

The theory of the Warré hive is that since the queen will lay eggs in the bottom comb, and the workers will store honey in the top comb, Warré hive users need to extract honey by removing the top-most box from the hive — the box under the quilt. This box is taken away and the honey extracted by cutting out the comb and pressing it to squeeze out the honey. A common method of doing this is the use of a mechanical fruit press which can be procured either online or through many large specialist kitchen shops.

HORIZONTAL TOP BAR HIVES

A popular alternative to the Warré hive is the Kenyan top bar hive and its variant, the Tanzanian top bar hive. These two hives are often called horizontal top bar hives (HTBH), since extra room for a growing colony is obtained by allowing the colony to expand horizontally not vertically as it does in either a Warré or a Langstroth hive.

In the sections on both the horizontal top bar hive and Warré hives, I will use the terms frame and top bar interchangeably. (The term frame really applies to Langstroth hives since it contains a top bar, two side bars, as well as a bottom bar. Top bars in both Warré and horizontal top bar hives may contain side bars but do not use bottom bars.)

The advantages of the Kenyan and Tanzanian horizontal top bar hives over the Langstroth hive are:

- The horizontal top bar hive is easy to make out of scrap wood or wood inexpensively purchased from a builders' supply store.
- No heavy lifting is required when inspecting or managing the hive.
- Frames can be easily removed for extraction.

- The horizontal top bar hive can be built so that the top is at waist level, simplifying inspection of the hive.
- Access to frames of brood or honey is straightforward, so they can conveniently be used as a training tool for new beekeepers. This is because both the brood chamber and the honey chamber (super) are at the same level and the beekeeper need not first remove the super to gain access to the brood chamber.
- A single horizontal top bar hive can be used to house two separate colonies by the use of an internal partition, usually called a follower board, suitably positioned to keep the two colonies apart.

The disadvantages of the horizontal top bar hive are:

- The hive does not use standardised parts like the Langstroth and a beekeeper may have difficulty purchasing a ready-made horizontal top bar hive or kit hive.
- The hive needs to be placed on a level surface and, due to its large footprint, this may be more difficult than for a Langstroth or Warré hive.
- Due to its larger size, moving a horizontal top bar hive may be more difficult than moving either a Langstroth or Warré hive. This can be overcome, however, by sliding the horizontal top bar hive onto a wheelbarrow which can be performed easily by one person.

Before discussing the management of a horizontal top bar hive the difference between Kenyan and Tanzanian top bar hives should be noted. The Kenyan horizontal top bar hive has sides that slope at about 30 degrees toward the base while the Tanzanian horizontal top bar hive has vertical sides. The remainder of the hive construction is the same except that the Tanzanian horizontal top bar hive does allow the use of frames with side bars if the beekeeper wishes to add them.

12.14: Making a hive out of a large log.

12.15: A top bar hive on a rural property in Victoria.

This is an advantage of the Tanzanian hive as frames similar to those used in Langstroth and Warré hives can be used. Side bars make a frame easier to remove as there will not generally be any comb attached to the sides of the hive. In addition side bars afford some strength and stability to the hanging comb, particularly in periods of very hot weather where melting from the top bar becomes a possibility.

A new colony may be introduced into a horizontal top bar hive with any unneeded space blocked off by the use of internal follower boards. As the colony grows and requires more room, the follower board can be moved outwards and more top bars added to the area housing the colony.

To enable the bees to enter and exit the hive, around one to three 25-millimetre diameter holes are drilled, usually on one of the hive walls. These entrances can be blocked with corks when not required. Suitable corks can be obtained from stores supplying home winemakers or brewers. Alternatively, used leftover wine corks can be washed and aired to remove odour.

Although the top bars provide a continuous surface over the top of the hive this surface contains cracks and is not weatherproof. To overcome this, a lid is placed over the entire horizontal top bar hive. This lid may contain extra insulation such as styrofoam or builder's blue house insulation sheeting to aid in further keeping the hive at a stable temperature.

Similar to the Warré hive, a horizontal top bar hive frame needs a thin strip of starter foundation attached to the top bar to guide the bees in drawing straight comb. This starter can be a thin strip of beeswax foundation inserted into a groove on the underside of the top bar frame held in place by pouring melted wax into a groove cut or routed for this purpose. An alternative method is to attach a triangular strip of wood to the underside of the top bar so that the bees will use the edge of the wood to guide their comb building. Popsicle sticks can also be glued into the routed grove to form a sharp edge to guide the bees in their comb making. If no guide is placed on the underside there is a risk that the bees will start drawing the comb at the edge of the top bar. This will result in an untidy, messy distribution of comb that will congest the hive and will prove difficult for the beekeeper to remove without damaging the honeycomb. This problem is true in particular with the Kenyan vertical top bar hive with its sloping sides, for as the bees build comb they attach it to the sides of the hive for strength. This can make removing frames difficult as there is the possibility that the entire comb can break off, leaving a mess for the beekeeper to clean up. To overcome this, a long sharp knife is needed to gently cut the comb from the hive walls before removing the top bar and, hopefully, the intact comb with it. If the comb is attached to the side walls this is usually only for the top 5 to 10 centimetres down the side of the box.

Of the three types of hive discussed in this book — Langstroth, horizontal top bar

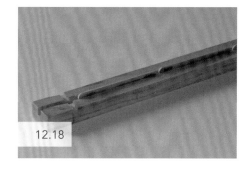

12.16: Inspecting a small top bar hive.

12.17: A strip of wax starter being built out by a colony of bees in a top bar hive.

12.18: The use of Popsicle sticks to give the bees an edge to start building comb from is a popular method of making frames for both Warré and Kenyan top bar hives.

hives and the Warré vertical top bar hive — I have had the practical experience of keeping my bees in primarily the Langstroth, which I admittedly prefer for my own convenience. I have, however, over the past two seasons run both a horizontal top bar hive and a Warré vertical top bar hive in order to educate myself on their management and to enable a more empirical assessment of the respective hives. It is also hoped that over a period of a few seasons as well as gaining some insight into the day-to-day practicalities of these types of hive I will observe any differences in bee populations hived in the different structures.

For the beekeeper interested in using one of the top bar hives there are three apparent disadvantages that may or may not prove of importance.

- A horizontal top bar hive is not cheap to buy as a kit or ready assembled. Top bar hives, however, are inexpensive and easy to

12.19

12.20

12.21

12.22

12.19: Removing a frame from a Kenyan top bar hive is sometimes difficult since the bees attach comb to both the frame and the sides of the hive. The easiest way to remove the frame is to use a long knife to cut the comb from the sides of the hive.

12.20: Step 1: Inspecting a frame from a Kenyan top bar hive. Since the foundation in a Kenyan top bar hive frame does not have any sides or wire to support it, if the frame were to be turned upside down the foundation would probably break off.

12.21: Step 2: To inspect the opposite side of the frame first rotate the frame by 90 degrees so that the top bar is vertical.

12.22: Step 3: Next twist the frame by 180 degrees so that you can inspect the other side of the foundation.

make, with plans readily available in books and off the Internet, for those with the necessary practical skills.

- Demand for and resale value of a top bar hive is a lot less than for a Langstroth hive.
- Both vertical top bar hives and horizontal top bar hives produce less honey than Langstroth hives. The reason for this is that since neither of the top bar hive designs uses queen excluders, the queen is free to lay eggs on the same frames as the workers store honey. This results in some frames of drawn comb holding both brood and honey. The honey is stored on the upper most part of the comb while the brood is on the lower part of the comb. This means that it is not possible to harvest the honey without killing many brood, which natural beekeepers would prefer not to do.

Both Warré and top bar hives are expensive and often difficult to purchase. As a result, many people apply natural beekeeping methods to the colony but buy a Langstroth hive in which to house their bees. This compromise is perfectly acceptable and is recognised as a legitimate approach to natural beekeeping.

SUMMARY

- Natural or sustainable beekeeping is gaining in popularity and there are now many proud owners of both Warré and Kenyan top bar hives. Natural beekeepers are often more concerned about the state of the environment and the plight of bees, in particular. The widespread planting of monocultures and the poor management practices that do not take into account the welfare of the bees are two of the concerns that natural beekeepers have about traditional beekeeping. Natural beekeepers view their relationship as a co-operative one with the bee and usually harvest less honey from a hive than the traditional Langstroth beekeeper. The honey taken is limited to the excess not required by the colony.
- The top bar hives mentioned above are usually difficult to purchase as a kit and if available they are expensive. This will no doubt change as the popularity of top bar hives continues to increase. In the meantime, beekeepers who adhere to many natural beekeeping practices successfully use the Langstroth hive which is both readily available in kit form and at a relatively low price.

13.

The bee-friendly garden

The interdependent relationship between the honey bee and the flowering plant is one which has developed over millennia. The bee relies on the flowering plant for nectar and pollen without which it could not survive. Pollen, which is produced and found on the male organ of the flower, provides protein for the bee and is an essential component of the bee bread fed to developing larvae. Nectar, a sweet sugary solution, supplies carbohydrates to the bee and is produced in special glands called nectaries generally found in the deepest part of the flower. In some plants additional nectaries are found on petals, leaves and stems.

To gain access to a flower's nectary the bee will brush past the pollen-laden anthers or male part of the flower. Some of the pollen will adhere to the hairs on the bee and while it continues foraging the bee will inadvertently transfer some of the grains onto the sticky stigma, part of the female organs of the flower. A plant may be self-pollinating which means it will reproduce using its own pollen transferred onto its own stigma or it may be of a type that requires cross pollination from another plant of the same species. Due to the bee's flower fidelity, or habit of visiting one particular type of flower during a foraging trip, many plants of the same species will be successfully pollinated and will reproduce.

BEE-FRIENDLY FLOWERING PLANTS

As a beekeeper it makes sense to populate the area around your hives with those plants that provide good supplies to your bees of the pollen and nectar essential for their wellbeing. There are myriad flowering plants available to suit the various climatic and gardening conditions found throughout a large continent like Australia. While advice can be sought from local

13.1: A hive is part of the garden picture.

garden centres and nurseries on the types of plant likely to thrive and flower in a certain area, for the beekeeper there are other additional considerations apart from personal preference that will influence plant choice. To earn a place in the bee-friendly garden a plant has to offer pollen, nectar or both and make these easily available to the visiting bee.

PLANT CHOICE

Whether you are renovating a small area of an existing garden or planning a new garden to maximise the forage available for your colonies of bees, there are certain initial considerations that will assist your final decision on what to plant. The following questions will guide you in your plant selection.

DOES THE PLANT FLOWER? WHEN, HOW OFTEN AND FOR HOW LONG?

While it is easy to quickly ascertain if a plant offers flowers, a further critical consideration before choosing a plant for the bee-friendly garden is the frequency of flowering. Find out at what time of the year a plant flowers, how often and for how long. Ascertain if it will repeat flower throughout the year or just the once. If it flowers once but has a lengthy flowering period it is worth considering. While for many plants flowering is restricted to the spring, during this time they produce copious amounts of pollen and/or nectar just when the hive is building brood and bee numbers are rapidly increasing. In my garden several large old-fashioned single-flowered climbing roses keep the bees working to capacity for a number of weeks. Due to their size and voluminous flowering their contribution, particularly of pollen to the garden's resident colonies, is important.

Many plants can be encouraged to repeat flower or sporadically flower for longer

13.2: Spring flowering climbing rose 'Pinkie' is a bee favourite in the author's garden.

13.3: Bluebells, one of the many bee attracting spring bulbs.

13.4: Summer flowering Crepe Myrtle.

13.5: Sedum provides summer flowering for hot dry conditions.

periods by judicious pruning and deadheading. Do a little research, check the plant label and see if some of the flowers you would like to grow will do this.

Initially limit yourself to a few good plant choices to start improving or developing your garden. Ensure that you select plant species that will offer a succession of flowers throughout the year and buy three or four plants of the same type, placing them together in blocks or informal groupings. As your knowledge grows and you become more confident with your plant choice you can add more of those species that do well in your particular area.

To create an environment that is more bee-friendly the aim is to choose a variety of plants that offer a succession of flowerings throughout the year. It is relatively easy to choose spring and early summer flowering species due to the large number of available plants that offer flowers during this period. Of equal importance are the late summer and autumn flowering plants that will sustain colonies after their initial build-up. Access to pollen and nectar is vital at this time as bees are building condition prior to winter and nutritious forage and continued storage of supplies is critical to the successful overwintering of colonies.

IS THERE ACCESS TO NECTARIES?

Often plants that succeed well in the garden and those most attractive to bees are the original species plant or old-fashioned cultivar that naturally evolved before we decided to improve on nature's design. The large double flowers so attractive to us with their complicated jumble of petals provide little or nothing in the way of pollen for the bee as they are often sterile; their sexual organs have disappeared or been modified to become extra petals. Nectaries, if they exist, are usually difficult or impossible for a bee to access.

Highly bee attracting are simple or single flowers with an open, shallow cup-like face, as both the nectar and pollen that the bee seeks is readily accessible. Many single-flowering roses, poppies, spring blossom and various daisy species are good examples of this shape and type of flower.

Bees collect nectar from flowers using their proboscis, or long tongue, that measures around 5.7 to 6.7 millimetres. A flower with a nectar-producing part any deeper than this will not be within the reach of the average honey bee although will appeal to many smaller native foragers.

A few plants with tubular or bell-shaped flowers may pose access problems for the bee as their nectaries are too deeply hidden at the base of a narrow tube of petals. For the most part, however, flowers with this shape are highly attractive to the bee and should be included in your plant choice. If you can look at the plant in the garden centre or plant nursery choose those tubular flowering plants that have a corolla tube (nectar-bearing part) wide enough for the bee to gain entry. Also a good choice are those tubular-shaped flowers that have overlapping petals which spread when the visiting bee forces them apart, often to completely disappear inside the flower intent on its nectar-seeking mission. Salvias, Grevilleas, Foxgloves, Penstemon and many fuchsia species (including some native types) have tubular-shaped flowers and are excellent bee-friendly choices if your garden conditions are suitable.

Plants with a stem or stalk ending in a cluster of tiny individual flowers are particularly bee attracting. In addition to their pollen and nectar rewards the visiting bee is able to conserve energy during a foraging trip with many flowers readily available to it on the same stem. Lavender is an excellent example of this type of multi-floral flower as are *Achillea, Buddleja, Callistemon, Echium* and the various *Hebe* species.

Another similar flower type usually found at the end of a long slender stem and highly attractive to your bees is the round globe or semi-spherical shaped flower. These flowers are also composed of many individual tiny flowers and again provide the bee with energy efficient forage. *Echinops* species, the globe artichoke (*Cynara scolymus*) and *Allium* species such as garlic, onions, chives and the beautiful but underused flowering garden alliums are examples of this shape.

In Tasmania, the bumblebee is found alongside the honey bee and while it no doubt provides competition for forage, it also often enables the honey bee to gain access to those plants with long tubular flowers and hidden nectaries that would otherwise be inaccessible and unavailable to them. A bumblebee is able to pierce a hole in the outside petals of a flower near or over the location of the flower's nectary. It then inserts its tongue, or proboscis, into the opening to obtain nectar from the outside of the bloom. Once opened up like this, other pollinators, including the honey bee, will take advantage of the easy access offered to the flower's nectar.

SIMPLE SINGLE FLOWERS

13.6: Almond blossom.

13.7: Poppy.

13.8: Euryops is a winter-flowering daisy.

TUBULAR SHAPED FLOWERS

13.9: Salvia.

13.10: A bee disappears inside a fuchsia flower.

13.9

13.10

BELL SHAPED FLOWERS

13.11: Blueberry flowers.

13.12: Comfrey.

MANY INDIVIDUAL FLOWERS ON ONE STEM

13.13: Hebe.

13.14: Californian Lilac.

13.14

13.15: *Buddleja davidii*, commonly called butterfly bush.

13.16: Garlic chive.

13.17: Common chive.

WILL THE PLANT THRIVE AND FLOWER IN THE AVAILABLE CONDITIONS?

Personal plant preference does to some extent become a secondary consideration in the bee-friendly garden. The primary focus of design and plantings is to offer year-round forage to the honey bee within the bounds of what will flower and survive in the conditions your property is able to provide.

As a starting point, survey your local area and look at the type of flowering plants that will survive and flourish — many do despite obvious neglect. Consider the overall climatic conditions of your area; frost tender plants will not flourish in very cold or snow prone parts of the country. Similarly, in hot dry arid areas heat-sensitive water-thirsty plants (often those with large fleshy leaves) will not thrive and you will waste a lot of time and energy trying to keep them alive.

Walk around your own property in the morning and again in the afternoon. You may need to do this several times over a period of time. Consider the effect of sunlight and of shade as this will help you decide on the choice and eventual position of your plants. Work out how much sun the area to be planted receives throughout the day and how this is affected by the changing slant of the sun as it alters during the various seasons. If, for example, you are considering a sun-loving species of flowering plant, many of these will require four to six hours of good sun each day. If they receive filtered sunlight or are in shade for long periods they will not thrive and will become weak and succumb to both pests and disease.

Give some thought to the aspect of your property and how this will affect your plantings. The area to be planted may face north but if there are tall mature trees, high fences or outbuildings these may affect the amount of sunlight reaching your plants for a large part of the day. Large trees will also provide both root and water competition for your establishing plants and inhibit their growth. If you live in an area prone to winter frosts bear in mind that buildings and dense shade can create 'frost pockets' for long periods in the winter and cause burning to sensitive plants from which they may not recover.

If your garden area is very exposed, for example on the top of a hill, a rise or near the coast and subject to salt-laden winds, these conditions should also be taken into account when you make your plant choices.

While you cannot do much about the aspect and climate of your property you can greatly improve soil condition and enhance the survival prospects of your chosen plants. Soil is a living, breathing medium and its overall state is critical to whether plants will thrive and mature to the point of flowering and providing optimum nutrition for your bees. All soil from the opposite extremes of sandy to hard clay can be greatly improved with the addition of organic compost. A good supply of organic material will condition the soil in your garden, increase its water holding properties, provide a home for millions of soil essential micro-organisms and attract and keep worms breeding and multiplying. Compost will also add the essential elements necessary to feed growing plants with the minerals and nutrients they require for healthy development and all-important flowering for your bees. A good layer of organic mulch will suppress weeds and provide a moisture-retentive protective blanket around your new plants.

13.18: 'Cottage Garden Rose' with its abundant flowering is a bee favourite in spring.

COLOUR AND GROUP PLANTINGS

Nature has endowed the bee with complicated compound eyes, or thousands of tiny lenses that are combined by the bee's brain into one picture. A group or several blocks

13.19: A sight that is difficult for a passing bee to miss.

13.20: What we see with our ordinary human vision.

13.21: What is seen by a bee with its ultra-violet vision.

or waves of plantings in bee-friendly colours will stand out to any passing bee and will be more obvious than a single specimen, unless of course the specimen is a large shrub or a tree covered in a multitude of flowers.

Bees are particularly drawn to the colours blue, purple, violet, white and yellow. During a foraging trip a bee tends to show an affiliation for one particular type of flower and will often ignore other nearby flowering plants that also provide pollen and nectar. Armed with this knowledge of bee behaviour you can plan a renovated or new garden by focusing on the colour and type of plant known to be favoured by the bee and then plant flowers of a single species in these colours in clumps or groups throughout your garden.

The bee has the ability to see ultra-violet light, which humans cannot, and many of the marks or features it finds attractive in a flower are not visible to us. While a bee cannot see the colour red many flowers of this colour have other visible markings such as black spots and ultra-violet lines that act as a map to guide it to the nectary of the flower.

By choosing groups of plants of differing sizes and height you will add structure and interest to your garden. In addition to being more visually appealing to you this will help your bees as different sizes and types of plant warm up in the sunshine at differing rates and as a result will release their nectar flow at different times of the day. Nectaries are not passive flower parts but are complicated active organs allowing many plants to regulate at what time of the day nectar will be secreted, in what concentration and to increase the flow in the presence of visiting pollinators.

Plan your garden so that you have three or four different types of plant flowering at any one time to aid your bees' nutrition. Concentrate on those plants that like a sunny position rather than a shaded spot in the garden; the bee favours plants in sunny parts of the garden and will visit flowers in these areas and often ignore those in shade.

FURTHER CONSIDERATIONS FOR THE BEE-FRIENDLY GARDEN

SPRAYS AND POWDERS

If you use proprietary sprays and powders to control unwanted insects and other garden pests consider the effects that these will have on your own bees and all other beneficial insects and pollinators. Bees are very susceptible to pesticides, herbicides, insecticides, fungicides and their residues. You can decimate or at least severely compromise the health of the adult bees or brood in your hive with one careless application of spray. Many sprays available these days are systemic and they are absorbed into the living tissues and transportation systems of the plant. Any pollen, nectar or water droplets on the plant will be taken back to the hive by a foraging bee as nectar or pollen, stored and eventually fed to the developing brood. Topical sprays for either pests or disease do not discriminate between beneficial and pest insects and will harm or kill them all. Dusting powders such as derris dust, often labelled as a natural product, are also harmful to foraging insects and to your bees. You will perhaps decide to live with a few spots, chewed leaves and many perceived plant problems of a cosmetic nature rather than risk compromising your own colonies and those of other beekeepers.

NATIVE VERSUS EXOTIC

The European honey bee has adapted well to the native flora of Australia and, in particular, to the copious nectar flow of many eucalypts. An increasing number of people prefer to plant native or indigenous species in their gardens and as the point of the bee-friendly garden is to provide flowering plants attractive to your bees for most of the year, with some research and the assistance of a specialist nursery you should be able to achieve a bee-friendly native garden with little difficulty in whatever region of Australia you live. For most of us a mixture of both native

13.22: Native hibiscus *Alyogyne huegelii*.

13.23: *Brachyscome multifida*, commonly called rock daisy.

13.24: *Hibbertia scandens*, or snake vine.

13.22

13.23

13.24

and introduced flowering plants will provide the largest choice of bee-attracting species for our gardens and ensure that our bees are easily able to access both pollen and nectar at various times throughout the year.

BEE-FRIENDLY NATIVE FLOWERING SPECIES

Banksia (*Banksia* spp.)

Brachychiton (*Brachychiton* spp.)

Bottle brush (*Callistemon* spp.)

Gum tree (*Eucalyptus* spp.)

Grevillea (*Grevillea* spp.)

Guinea Flower (*Hibbertia* spp.)

Native daisy bush (*Olearia* spp.)

Paperbark, Honeymyrtle, Tea tree (*Melaleuca* spp.) — note that some individual plants of the *Melaleuca* spp. are commonly called tea tree; examples are *M. minutifolia* and *M. alternifolia*.

Senna (*Senna* spp.)

Tea tree (*Leptospermum* spp.)

Do your own research or ask a specialist native nursery or your local garden centre to suggest suitable plants within each species for your area and requirements.

INCORPORATING BEE-FRIENDLY PLANTS INTO THE GARDEN

Regardless of the size of your garden you can, depending on your inclination and time constraints, add a considerable amount of forage for your own and other visiting bees. While you may not wish to drastically alter a garden that you have nurtured and developed over many years it is always possible to find areas that you can add to or renovate with bee-attracting flowering plants.

HERB GARDENS

If you are not really a gardener but can find a small amount of sunny space, consider planting a herb garden which will add to your own health and to the nutrition of visiting bees. A herb garden can be elaborately designed as an attractive addition to your property or be a simple patch of sunny fertile soil planted with a few herbs. Plant a flowering border or low hedge around the chosen area using thyme, garlic chives, sage or a low-growing species of lavender (*L. stoechas*) and you will quickly add interest to your garden and extra flowers for your bees.

A large number of the commonly used kitchen herbs such as rosemary, the various thymes, sage, marjoram, oregano, basil and parsley are very attractive to your bees.

It will, of course, be necessary to allow your herbs to flower before they are useful to your colonies and this is not difficult if you limit your gathering to pinching out various areas of the plant for your own kitchen use. This will have the added effect of the plant growing further branched stems to support more flowers for your bees.

HERBS FOR BEES AND YOU

Sweet basil (*Ocimum basilicum*)
Borage (*Borago officinalis*)
Chives (*Allium schoenoprasum*)
Comfrey (*Symphytum* spp.)
Lavender (*Lavandula* spp.)
Lemon balm (*Melissa officinalis*)
Marjoram (*Origanum majorana*)
Mint (*Mentha* spp.)
Oregano (*Origanum vulgare*)
Parsley (*Petroselinum crispum*)
Rosemary (*Rosmarinus officinalis*)
Sage (*Salvia officinalis*)
Thyme (*Thymus* spp.)

13.25

Many herbs are vigorous. Comfrey needs space and is not a choice for a small garden. The mint family in particular should be planted with care and contained.

VEGETABLE GARDEN

If you prefer the edible type of garden ensure that you leave several plants for your bees by allowing them to flower. Brassicas such as cauliflower, broccoli, cabbage and Brussels sprouts offer both pollen and nectar when allowed to flower, whereas many modern cultivated vegetables and self-pollinating peas and beans will offer nothing at all to your bees.

13.26

13.27

13.25: Rosemary.

13.26: Borage.

13.27: Lavender.

13.28

13.29

13.30

Strict soldier-like rows of vegetable plantings can be broken up by interspersing with groups of flowering plants which, in addition to providing nutrition for passing bees, may also serve to confuse and divert some of the pest species attracted to your vegetables.

There is a good deal of information available on both companion planting in the vegetable garden and planting to repel pest species. Further investigation of flowering species with these extra possibilities is a winning proposition for both your bees and you as a gardener.

Consider planting a green manure crop, which will grow quickly, flower for your bees and before it sets seed can be dug to provide humus and minerals for your soil. Buckwheat is a plant that offers much; it is highly attractive to bees, produces edible seed for you to eat and can finally be dug in to fertilise and add humus to your soil as a green manure crop.

FRUITING TREES AND PLANTS SUITABLE FOR BEES

Apple, crab apple (*Malus* spp.)

Avocado (*Persea americana*)

Blueberry (*Vaccinium corymbosum*)

Blackcurrant (*Ribes nigrum*)

Banana (*Musa* spp.)

Broccoli, cabbage, cauliflower, brussels sprouts (*Brassica* spp.) — leave them to flower

Cucumbers (*Cucumis sativus*)

Lemon, lime, orange, mandarin, cumquat (*Citrus* spp.)

Macadamia (*Macadamia integrifolia* var.)

Passion fruit (*Passiflora edulis*)

Persimmon (*Diospyros* spp.)

Plum, cherry, peaches, nectarines and almonds (*Prunus* spp.)

Pumpkin (*Cucurbita maxima*)

13.28: Greek basil.

13.29: Chilli plant.

13.30: Leave a few vegetables to flower for your bees to forage in.

Raspberry (*Rubus idaeus*)
Redcurrant (*Ribes rubrum*)
Strawberry (*Fragaria x ananassa*)
Tomato (*Solanum lycopersicum*)

VERTICAL FLOWERING SPACE

HEDGING

Instead of the overdone non-flowering box hedge choose a low flowering plant that also responds to clipping if this is the look you prefer. You will have to restrain the clippers until after the flowering period of the particular plant has passed but your efforts will often result in a further flush of flowers for your bees. Taller hedging can be used to create windbreaks and soften existing fence lines. *Abelia, Callistemon, Ceanothus, Escallonia, Eriostemon, Westringia, Hebe* and the rosemary leafed Grevilleas are all species that respond well to hedging and are good examples of bee-attracting shrubs.

BEE-ATTRACTING FLOWERING SPECIES FOR HEDGING

Abelia (*Abelia* spp.)
Escallonia (*Escallonia* spp.)
Grevillea (*Grevillea* spp.) — fine rosemary leaved types are best
Hebe (*Hebe* spp.) — numerous cultivars in various sizes and leaf types
Lilly Pilly (*Syzygium* spp.)
Native rosemary (*Westringia* spp.)
Rugosa Rose (Rosa rugose)
Viburnum (*Viburnum* spp.) — deciduous and evergreen types available

CLIMBING PLANTS

Flowering shrubs or vines can be planted against existing fencing and walls to increase the vertical flowering space of your garden and the forage available to your bees. Pergolas and arbours add special interest to any property and are ideal supports for large flowering vines and climbers like the vigorous Wonga vine, *Hardenbergia*, clematis and climbing rose species.

Selection of climbing roses should concentrate on old-fashioned or more recent cultivars with either single or loosely formed blooms held in clusters on the plant. Avoid hybrid tea roses; their blooms are usually inaccessible to the bee until nearly spent and they do not produce sufficient numbers of flowers to be useful for forage. Consider the tough Rugosa roses. These are commonly used in other parts of the world for hedging and many have numerous large single flowers that are very

13.31: Hebe 'Lavender Blue' used as a loose hedge.

13.32: Close up of Hebe 'Lavender Blue'.

13.33: San Pedro cactus.

13.34: *Rugosa* 'Frau Dagmar Hartopp'.

13.35: Clematis 'Marie Boisselot'.

13.36: Clematis 'Nellie Moser'.

attractive to bees. They are disease resistant and most have colourful hips for birds in the autumn. Unfortunately, they also have a number of thorns that may bother you — but not your bees.

Climbing plants can grow to an enormous size so choose your species wisely and do not plant anything that will become too large and heavy for the structure supporting it.

Provided with the right support, in temperate parts of the country large clematis are easy to grow. Several varieties can be planted close to each other to provide flowers from spring through to the early part of summer. Most will spot flower again in the autumn.

GROUND COVERS

Suitably chosen low-flowering groundcover plants can be used in your garden both as living mulch and as an extra source of forage for your bees. Some of the popular varieties such as the pink or white native *Myoporum parvifolium* (Creeping Boobialla) and mauve or purple *Scaevola aemula* (Fan flower) are easily grown. The exotic purple or white *Bacopa* (Sutra) species and the mauve flowering *Convolvulus mauritanicus* are also attractive useful ground covers which when planted should be kept in check as they can become rampant in some areas. Prostrate ground-covering Grevilleas provide considerable pollen and nectar during the spring build-up of the hive and there are numerous original species and cultivars available to suit most garden environments.

13.37

13.37: Clematis 'The President'.

GROUND COVER PLANTS SUITABLE FOR BEES

Creeping Boobialla (*Myoporum parvifolium* spp.)

Campanula (*Campanula poscharskyana* spp.)

Dead nettle (*Lamium* spp.)

Fan flower (*Scaevola aemula* spp.)

Grevillea (*Grevillea* spp.) — and also prostrate forms

Guinea Flower (*Hibbertia* spp.)

Pig face (*Carpobrotus glaucescens* spp.)

Sturt's Desert Pea (*Clianthus formosus* spp.) — grafted varieties are less susceptible to root rot

Thyme (*Thymus* spp.)

GARDENING IN POTS

If you live in an apartment or dense urban situation with or without your own colony of bees you can add to the flowering diversity of your area with potted flowers, herbs and some smaller flowering shrubs. The addition of several well-chosen bee-friendly flowering plants to your balcony or courtyard will offer food for visiting bees and will add interest and colour to your immediate surroundings. While you may feel your contribution is small you will add to the overall forage available for bees and other pollinators within your wider neighbourhood. Any flowering plants, even one or two, are better than none at all.

ENVIRONMENTAL WEEDS

It is perhaps obvious that no one should plant and nurture anything that has been declared an environmental weed or is likely to become one. It is not always clear which plants fall into this pest category and to confuse matters further what is a weed in one part of Australia may not be in another. Your local council or plant nursery should be able to provide advice on what not to plant and many councils and local authorities supply residents with a list of plants considered environmental weeds and not safe to plant. Declared weed species may be native to Australia or introduced. Many environmental weeds are former garden plants or native species that have adapted too well to the conditions offered in a particular area, to the point that they threaten to take over large areas if left unchecked. Many of these species are very attractive to bees and in more rural areas are known to provide valuable forage during times of eucalypt and other flowering shortages. In the domestic situation the dandelion and many common flat-leaved lawn 'weeds' are sought out by the bee and you may decide to leave them where they are for this reason alone.

13.38: Rose 'Crépuscule'.

13.39: Star Jasmine.

13.40: Even one pot of bee-friendly flowers is better than none at all.

WATER

As a general rule avoid flowering plants that like a lot of water. An increasing trend on plant labels is an indication of how much water a plant requires via a diagrammatic water drop system. A plant label with one drop indicates low water requirements. The more drops a label displays the greater the plant's water requirements for optimum growth. All plants will initially require some watering to establish their roots, but unless you live in a particularly dry or arid region, once established many can survive on natural rainfall alone and should be favoured choices.

If you live in an area prone to long dry spells consider concentrating on flowering shrubs and trees, as generally a shrub or tree is better able to sink its roots deep enough to survive short periods of hot weather and drought and will require less overall water and care. Nevertheless, there are many hardy perennials that will survive short dry periods and will both provide beauty to the garden and food and nutrition for bees and other foragers. Perennials are plants that do not need to be replanted annually; they have the added advantage that many species are easily propagated by taking cuttings from existing plant stocks or from plants in the gardens of others.

If you have the room consider adding a garden pond, even a small one, to aid your bees in their search for water. Ensure that you plant this with a variety of marginal and aquatic species that can be used by your bees as safe landing platforms to access the water while they drink.

13.41

13.41: Cotoneaster is an environmental weed in many parts of Australia but it is a magnet for bees in the spring and in autumn its red berries are loved by birds (who defecate its seeds ensuring its propagation).

A GARDEN IS A WORK IN PROGRESS

However you decide to improve your property and garden for your bees you will make mistakes or change your mind about some of your initial choices. A plant that does well in another beekeeper's garden may not thrive in yours. Any garden, including one being renovated or designed to be bee-kind, is never a static picture but an evolving one requiring ongoing labour, so if you are unhappy with a plant choice or its placement you can always change or improve on your efforts over the years. This becomes more feasible and an inexpensive task once you successfully take your own cuttings from those bee-friendly plants recommended by other beekeepers.

If you have a large area to fill consider planting a few flowering 'sacrifice' plants to fill in the spaces until your chosen specimens gain enough bulk to provide useful forage. You may even decide that these 'fillers' provide more flowers for your bees than your chosen specimens and decide to prune and leave them.

Sow your own annuals, many of which can be purchased as seedlings quite inexpensively. If you would like to experiment with seeds start with sunflowers (*Helianthus annuus*) or *Limnanthes douglasii*, the poached egg plant. Both have larger seeds that are easy to handle and plant. The poached egg plant makes a beautiful spring annual border or potted display and is highly attractive to bees offering them both nectar and pollen.

Avoid over hybridised plants (many will not produce viable seed) and limit your selection of annuals to the old-fashioned cottage garden plants such as alyssum, cosmos, single calendula, candytuft, nasturtiums, salvias and the native daisy *Brachyscome iberidifolia*.

Annual poppies are bee magnets with their copious amounts of pollen. Most varieties are easy to grow from seed but resent being moved so need to be sown where they are to flower. Although poppies readily self-seed it is a good idea to save a few seed pods either to share with other beekeepers or to cast around the garden the following spring to increase the numbers of flowers for your bees.

To be useful to your bees your plants must thrive and survive until the point of flowering. Even the most basic of composting efforts will improve the soil over time and improve the health and longevity of your chosen plants and as a result the nutrition of your bees.

While your efforts may be aimed at providing additional nutrition for your own bees you will, by adding more floral content to your garden, add to the wider biodiversity of your local area and attract other beneficial insects and native pollinators to make homes in your garden. Australia has many native social and solitary bees that can be encouraged to take up residence along with your hives of European honey bees. As you learn more about the flowers that are attractive to your bees and whether they are good sources of pollen or nectar you may find your interest in flowering plants will grow and become an extension of your expertise as a beekeeper.

Honey bee nutrition will become even more critical if the incursion of the *Varroa* mite spreads across Australia. The bee will have no initial resistance to the mite and will be affected by the many viruses it transmits. The hobby beekeeper will have an important role to play in both keeping and sustaining colonies of bees during the initial incursion. Providing additional forage from healthy plants during this period will benefit and support the bee, as it needs a wide variety of pollens to supply the complete source of the protein it needs and varied nectars to provide carbohydrates and energy.

While a single property or a garden cannot provide all the nectar and pollen requirements of a colony it does add to the variety of plants and nutrition on offer and contributes to and increases the overall diversity of the available flora within a particular area for all visiting bees and foragers.

Overleaf is a list of a range of plants that are bee-friendly and will bring bees into your garden.

13.42

13.42: European honey bee and native Blue-banded bee foraging together.

BEE-ATTRACTING ANNUALS

Alyssum (*Lobularia maritima*) — a perennial that is usually grown as an annual

Candytuft (*Iberis amara*) — a perennial form (*I. sempervirens*) is also available

Cosmos (*Cosmos bipinnatus*)

Forget Me Not (*Myosotis alpestris*)

Livingstone daisy (*Mesembryanthemum*)

Nasturtium (*Tropaeolum majus*)

Poppies (*Papaver* spp.) — adored by bees for their copious pollen

Poached Egg Plant (*Limnanthes douglasii*)

Sunflower (*Helianthus annuus*)

Swan River Daisy (*Brachyscome iberidifolia*)

BEE-ATTRACTING PERENNIALS

Bacopa (*Sutera cordata*)

Blue mist plant (*Caryopteris x Clandonensis*)

Catmint (*Nepeta* spp.)

Clematis (*Clematis* spp.) — both native and introduced

Elephants Ears (*Bergenia* spp.)

Fan Flower (*Scaevola aemula*)

Hebe (*Hebe* spp.)

Michalmas daisy (*Aster x frikartii*)

Pentas (*Pentas lanceolata*)

Penstemon (*Penstemon* spp.)

Pride of Maderia (*Echium candicans*) — plant with care and remove dead
 flowerheads as this can be invasive

Salvia (*Salvia* spp.) — annual and perennial types

Scabiosa (*Scabiosa* spp.) — especially *S. caucasica* and *S. atropurpurea*

Sea Lavender (*Limonium perezii*)

Sedum (*Sedum* spp.)

Thrift (*Armeria* spp.)

Valerian (*Centranthus ruber*)

Verbascum (*Verbascumbombyciferum*) — this is a bee magnet

Wallflower (*Erysimum* cvrs)

Winter rose (*Helleborus* spp.)

Yarrow (*Achillea* spp.)

BEE-FRIENDLY SHRUBS

Abelia (*Abelia x grandiflora*)

Banksia (*Banksia ericifolia*), Heath Banksia, Banksia 'Giant Candles', Saw Banksia

Buddleja (*Buddleja* spp.) — *B. alternifolia, B. salviifolia, B. davidii*

Californian Lilac (*Ceanothus* spp.) — both tall and compact varieties

Bottle Brush (*Callistemon* spp.) — upright, weeping and dwarf varieties

Escallonia (*Escallonia* spp.)

Grevillea (*Grevillea* spp.) — various heights and forms including prostrate

Hebe (*Hebe* spp.) — miniature, small and tall varieties

Hibiscus (*Hibiscus* spp.)

Mexican Orange Blossom (*Choiysa ternata*)

Native hibiscus (*Alyogyne huegelii*)

Native Rosemary (*Westringia* spp.)

Native Daisy Bushes (*Olearia* spp.)

Rice Bush (*Pimelea* spp.)

Rock Rose (*Cistus* spp.)

Rose (*Rosa* spp.) — with multiple single or loose-petalled blooms that open to show central cluster of stamens

Sticky Hopbush (*Dodonaea viscosa*)

Thryptomene (*Thryptomene* spp.)

Viburnum (*Viburnum* spp.)

Wax Flower (*Philotheca myoporiodes*)

FLOWERING TREES SUITABLE FOR BEES

Trees should be chosen and planted with a view to their eventual size. Many eucalypt and fruiting trees are available grafted onto smaller or dwarf rootstock, making them a good choice in a large range of garden situations.

Blueberry Ash (*Elaeocarpus reticulates*)

Jacaranda (*Jacaranda mimosifolia*)

Pin Cushion Hakea (*Hakea laurina*)

Silky Oak (*Grevillea robusta*)

Small Leaved Lilly Pilly (*Syzygium luehmannii*)

Water gum (*Tristaniopsis laurina*)

Native bees

There are two general types of bee found in Australia: social bees and solitary bees. Social bees, like the European honey bee and the native social bees, form colonies consisting of thousands of individuals that work together to ensure the survival and wellbeing of the colony. Social native bees are stingless and are often referred to by the common name Stingless Bees. In this chapter, to ensure that there is no confusion between the European honey bee and native social (stingless) bees, both of which are 'social', I will refer to the native social bee using the name 'social bee'. Solitary bees, as their name suggests, lead solitary lives, independent of other bees. As solitary native bees have no large food stores in their nests to protect they are usually not aggressive but they are able to sting and should be treated with respect. While female solitary bees live in prepared nests, male solitary bees usually congregate or 'roost' at night in the open, clustered together.

SOCIAL BEES

In Australia, there are 1600 described species of native bee, fifteen of which are social. These fifteen species can be classed into two genera (plural of genus): *Tetragonula* (formerly *Trigona*) and *Austroplebeia*. Only four species have been 'domesticated' and kept in hives. These are:

- *Tetragonula*
  - *Tetragonula carbonaria*
  - *Tetragonula hockingsi*
- *Austroplebeia*

14.2

- *Austroplebeia australis*
- *Austroplebeia symei*

The natural range of Australia's social bees is mostly in the warmer northern regions of the country. There are, however, differences between the geographic regions inhabited by *Tetragonula* and *Austroplebeia*. *Tetragonula* prefer the wetter coastal regions while *Austroplebeia* favour the drier inland regions. In order to survive in such different environments Australia's social bees have evolved a range of diverse behaviours.

Social bees have been an important part of indigenous culture for thousands of years with indigenous people using all of the nest contents, eating the honey, brood and pollen. In addition, the resin was used in crafts to waterproof baskets, and to attach tool and weapon heads to handles. The strong, tangy honey from social bees is called 'sugarbag' and commands a much higher price than honey made by the European honey bee. This difference in price is mainly due to the difficulty in collecting any marketable quantity of 'sugarbag' honey.

Australian social bees are small and black, measuring approximately 4 millimetres. *Austroplebeia* species can be distinguished from *Tetragonula* by their creamy coloured thoracic markings and nest architecture.

Australia's social bees have much smaller colony populations than honey bee colonies (10,000 bees vs. 50,000 bees), therefore they store proportionately smaller amounts of honey within their nests. Due to this fact, robbing a social bee nest may mean the death of the colony. This is particularly so in areas that are marginal for optimal foraging, such as the more southern regions of Australia. During a good nectar flow however, social bee colonies can store larger amounts of honey.

For *Tetragonula carbonaria* to fly, the ambient temperature must be at least 18° Celsius,

while *Austroplebeia* needs an ambient temperature of greater than 20° Celsius.

As with the honey bee, there are three castes within a native social colony. These are a single queen, a large number of female workers and a small number of male drones. Unlike the multiple mating queens of the European honey bee, social bee queens are monandrous, mating with only a single drone.

AUSTROPLEBEIA

There are nine species of *Austroplebeia* listed in *The Zoological Catalogue of Australia* and the two most commonly domesticated and studied are *Austroplebeia australis* and *Austroplebeia symei*. Classification of the different *Austroplebeia* species is based mainly on thoracic markings. *Austroplebeia* bees are generally black with varying levels of cream or yellow markings on the rear edge of the thorax (scutellum), along the thoracic margins and on the face. *Austroplebeia* are between 3.5 millimetres and 4.5 millimetres in length.

Austroplebeia live in some of the most arid regions of Australia, where the climate extremes are harsh and food resources are often scarce. Species within the *Austroplebeia* genus can be found throughout northern Australia, with *Austroplebeia symei* having the widest distribution. *Austroplebeia symei* can be found along the south-east coast of Queensland as well as in the northern parts of the Northern Territory. *Austroplebeia australis* is found in Queensland, in New South Wales as far south as Kempsey and as far inland as Inverell.

Austroplebeia choose hollow trees to build nests. *Austroplebeia australis* is typically found in hollows with a diameter of between 50 millimetres and 110 millimetres, and some nests have been recorded as being up to 6 metres in length. Brood populations may vary between 2000 and 13,000 and these variations may be due to cavity size, time of year and available floral resources.

All *Austroplebeia* construct spherical brood cells and, with the single exception of *Austroplebeia cincta*, make simple cell

14.3: *Austroplebeia* nest.

14.4: *Austroplebeia*.

14.5: Head of an *Austroplebeia* bee.

14.6: *Austroplebeia* defending its nest against a much larger *Apis mellifera*.

clusters. Open cells face outwards from the leading edge of the cluster. Clustered brood cells can be constructed to fit into the narrow, irregular cavities of the smaller trees or larger limbs favoured by *Austroplebeia*.

TETRAGONULA

Tetragonula were reclassified from the genus *Trigona* in 2012. Identification of species of the *Tetragonula* genus is often difficult as some members may vary in size depending on their location. The largest, *Tetragonula hockingsi*, measures approximately 4.5 millimetres in length, while the smallest, *Tetragonula clypearis*, measures only 3.5 millimetres in length. As a result, the different nest architectures are an important aid in determining an identification of the species.

The distribution of *Tetragonula carbonaria* is throughout northern and eastern Australia and extends from the Atherton Tablelands in Queensland to as far south as Bega in New South Wales, the most southerly geographic range of the social bees. *Tetragonula carbonaria* chooses large cavities in which to build its nest. Apart from trees, *Tetragonula* species also build nests in utility meter boxes, stone walls, beneath concrete paths and within door and wall cavities.

All of the *Tetragonula* species in Australia build elongated, vertically oriented brood cells in regular, or nearly regular, structures. As previously mentioned there are some differences in the way that the species build their respective nests that can aid in their identification. A description of these differences is beyond the scope of this book and interested readers can consult the many excellent native bee resources listed in the bibliography.

The amount of honey that can be harvested from a *Tetragonula* hive varies significantly. In Queensland and northern New South Wales, where bees can forage all year round, beekeepers are able to harvest 1 kilogram per hive per year. In the cooler regions further south, beekeepers should only harvest every two to three years, and it is recommended that hives kept in the Sydney basin or further south should not be harvested for honey at all.

14.7: *Tetragonula carbonaria* nest.

14.8: *Tetragonula hockingsi* nest.

14.9/14.10/14.11: *Tetragonula.*

ALTERNATIVE POLLINATORS

Pollination of crops using social bees is rare and commercial crops rely heavily on the European honey bee for pollination. Free pollination by feral honey bees is more widespread in Australia than in other regions of the world such as North America, Europe and New Zealand. Feral honey bees in these overseas regions have been almost wiped out by *Varroa* infestations and the resulting spread of diseases by the mite. The reliability of feral honey bee pollination services in Australia is also expected to be drastically reduced if *Varroa* is introduced into this country. One recent study counted 40 to 150 feral colonies per square kilometre of the European honey bee, *Apis mellifera*, in the Wyperfeld National Park in north-east Victoria. As an example of how an introduced pest can affect feral colony numbers, between 2002 and 2006 it is estimated that more than 4500 colonies of European honey bees died out in Australia as a result of infestation with the Small Hive Beetle, *Aethina tumida*.

The ongoing threat of disease to the European honey bee has caused many people to look for alternative pollinators including Australia's social bees. Interest in the use of social bees to pollinate macadamia trees increased when it was observed that the yield of macadamia nuts increased when they were planted next to uncleared native vegetation where there were significant numbers of native social bees. This is perhaps not surprising as both these native species of plant and insect would have evolved together over many millennia.

It has been estimated that Australian social bees have a flight range of 500 metres, although many beekeepers believe the range is more likely to be a kilometre. This is advantageous to the crop grower since any social bees brought in to pollinate crops are unlikely to forage further away to find other floral resources. Along with actual hive placement, the number of hives is important as fifteen to twenty native social bee hives per hectare are required in comparison with seven hives per hectare for the European honey bee. In addition to macadamia, social bees are also effective pollinators of lychee, avocados and watermelons.

Native social bees seem to be relatively free of disease and at the point of writing there are no reports of any brood diseases having been seen. They do, however, suffer from predation, parasitism, and colony infestation. They face a similar range of threats from pests as do the European honey bee. As an example, the Small Hive Beetle, *Aethina tumida*, can devastate both a weakened social bee colony as well as a European honey bee colony. Both *Tetragonula* and *Austroplebeia*, however, are more effective than the European honey bee at defending against Small Hive Beetle but can succumb if the colony is weak. Cane toads, *Rhinella marina* (formerly *Bufo marinus*), are also a threat to both European and social bee colonies.

Nest defence by native social bees is reported widely across the world. A variety of methods are used by native social bees to defend colonies including biting, resin daubing, chemical repellants, and locking onto the wings or body of intruders with their mandibles.

Tetragonula carbonaria exhibit a collective mechanism known as 'fighting swarms' during which bees from one hive attack or try to enter another hive. During a fighting swarm hundreds to thousands of these native bees are locked in aerial combat and fight to the death. These battles can last less than a day or extend over several days, restarting again each morning, and may result in thousands of dead and dying bees lying on the ground.

Many people in northerly regions of Australia keep social bees in hives, usually as an aid to the bee's conservation, as the amount of honey that can be harvested is often small. In the southern parts of Australia where social bees are not naturally found, such as Victoria, Tasmania or South Australia, keeping *Tetragonula* or *Austroplebeia* bees in hives is not recommended due to the colder climatic conditions. Solitary bees, however, are found across Australia and there is much that an interested person can do to attract and encourage solitary bees to nest in their garden. Details on attracting and keeping native bees can be found in the bibliography.

SOLITARY BEES

This section has been written by Erica Siegel, wildlife photographer and native fauna enthusiast.

There is a very large number of solitary bee species, ranging from minute bees, less than 3 millimetres long, to giants as long as 28 millimetres. Some of the more commonly seen and recognisable groups are described below.

BLUE-BANDED BEE — *AMEGILLA*

There are about eleven species of Blue-banded bees in Australia. They are found in every state including Tasmania and are one of Australia's long-tongued bees within the *Apidae* family and are known as buzz pollinators. These blue-banded bees range from 8 to 14 millimetres long; they have a sting but are not aggressive. They have thick red-brown hair on the thorax and glittering bands of iridescent, metallic blue or whitish hair across their black abdomens. Some, however, have green or orange bands. Males have five bands and females four. The colours are caused by microscopic tubes in each hair with the reflecting light causing the glittering colours.

They have yellow, cream or white markings on their faces and are often seen darting around the flowers of lavenders, and abelias and many other exotic and native flowers. The pollen is carried on their hind legs in sets of specialised hairs (termed scopae). The females build nests in shallow burrows in the ground, favouring river and dam banks, but they may also nest in mud bricks, soft sandstone banks or in soft mortar. Each female builds her own nest burrow but many females often nest together in the one place. The female Blue-banded bee constructs oval-shaped cells in her burrow, lining them with waterproof secretions. Before laying an egg in the cell she stocks it with a paste of pollen and nectar. After the egg is deposited she seals the cell with an earthen

14.12

14.13

14.14

14.15

cap and when all the cells are filled and capped, she covers the nest with layers of soil to close it.

According to J.C. Cardale (Australian National Insect Collection, Canberra 1968), Blue-banded bees live around 40 days and about three generations of bees hatch during summer, depending on the location. Baby bees take about seven weeks to hatch. Those that do not hatch overwinter in their sealed cells as prepupae, emerging in the next spring.

The nests of female Blue-banded bees are stalked by the Neon Cuckoo bee (*Thyreus nitidulus*).

Male Blue-banded bees roost in small groups out in the open at night, hanging on to twigs or stems with their mandibles and tucking their legs under their bodies. Recent research has shown that Blue-banded bees could be valuable pollinators of greenhouse tomatoes.

14.12: Female Blue-banded bee, *Amegilla cingulata*.

14.13: Blue-banded bee feeding on basil.

14.14: Female Blue-banded bee, *Amegilla cingulata*.

14.15: Male Blue-banded bees roosting, *Amegilla cingulata*.

14.16: Teddy Bear bee, *Amegilla (Asaropoda) bombiformis.*

14.17: Teddy Bear bee, female.

14.18: Male Teddy Bear bees roosting.

TEDDY BEAR BEE — *AMEGILLA (ASAROPODA)*

There are about 25 species of *Amegilla (Asaropoda)* recorded in Australia. Teddy Bear bees occur in all states of mainland Australia.

Most species of these rotund, furry brown bees are 7 to 20 millimetres long and are covered in dense red-brown hair but some are a lighter buff colour. They are one of Australia's long-tongued bee groups and the females carry the pollen on special thick hair on their hind legs. The largest, the Dawson's Burrowing bee, *Amegilla (Asaropoda) dawsoni*, is nearly 20 millimetres long and lives in Western Australia. It nests in groups of up to 10,000 in arid clay pans and mud flats. The males fight to their death to mate with virgin females when they emerge from their burrows as documented in the BBC TV series *Life*.

Other Teddy Bear bees build shallow 10-centimetre nest burrows in soft soil, creek banks, in weep holes of retaining walls, in gaps in sleepers surrounding garden beds and sometimes underneath houses. Many females may nest together in the one location but each female builds her own nest containing a cluster of 2-centimetre urn-shaped cells made of mud and lined with waterproof secretions. The cells all face upwards and the female stocks each cell with a paste of nectar and pollen on which she lays her egg. The cell is then sealed with a cap. Fully grown larvae may lie dormant until spring if winter is approaching.

Female Teddy Bear bees are stalked by the Domino Cuckoo bee, *Thyreus lugubris*, which will lay her egg in unattended nests. The Cuckoo bee larvae will consume the food meant for the larvae of the Teddy Bear bee.

Teddy Bear bees are known as buzz pollinators allowing them to pollinate flowers with enclosed pollen like *Solanum* sp., such as tomato, eggplant, chilli. They grasp the flowers with their front legs and use their flight muscles to vigorously vibrate the flower, dislodging the pollen.

They are also valuable pollinators of native wildflowers and forage on native flowers like *Senna clavigera, Hibbertia scandens, Hibbertia dentata, Dianella* sp., *Melastoma affine*, and exotic *Buddleja davidii, Cuphea* sp.

The males roost together at night hanging onto stems or twigs with their mandibles. They sometimes roost near the roost of Blue-banded bees.

GREAT CARPENTER BEE, YELLOW AND BLACK CARPENTER BEE — *XYLOCOPA (KOPTORTOSOMA)*

Great Carpenter bees are the largest native bees in Australia, ranging from 15 to 26 millimetres. According to measurements in Remko Leijs' revision of the Carpenter bee genus, the males are either the same size or slightly bigger than the females.

There are six species of Great Carpenter bees in Australia and they are found in the warmer climate of northern Western Australia, northern New South Wales, Queensland and the Northern Territory. The females have a glossy black abdomen and bright yellow hair on the thorax. Males are covered uniformly with yellow, brown or olive green hair. Both males and females have black wings.

They are called Carpenter bees because they cut nest burrows with their jaws in soft timber such as the dead limbs of mango, frangipani and jacaranda trees. Other known nesting trees include *Ficus* species, *Casuarina, Banksia, Lophostemon grandiflorus, Leptospermum* species, dead flower stalks of Grass trees (*Xanthorrhoea* sp.), Soursop (*Annona muricata*), dead saplings of *Eucalyptus robusta*, even old timbers in the backyard.

The nesting burrows may contain several tunnels and are partitioned into brood cells. The female lays her egg on a mixture of pollen and nectar then seals the cell with a disc of chewed wood particles before placing another egg on a paste of nectar and pollen, sealing it off till the tunnel is filled.

14.19

14.20

14.21

14.19/14.20: Great Carpenter bee, *Xylocopa* (*Koptortosoma*).

14.21: Great Carpenter bee, female.

New generations of females often use old nesting burrows. Female Great Carpenter bees are known to share the burrow with their daughters creating a small community, even sometimes feeding their adult daughters, and are therefore referred to as 'para-social'.

The Great Carpenter bees are attracted to native *Melastome affine*, *Senna*, foam bark (*Jagera pseudorhus*), *Canavalia rosae* and according to Dr Remko Leijs (Flinders University, SA) *Crotalaria*, *Wistaria* (*Callerya megasperma*) and exotic *Cassia fistula*. They also visit other exotic plants like *Grewia occidentalis*, Pigeon Pea (*Cajanus cajan*), Cashew tree (*Anacardium occidentale*), Leopard tree (*Caesalpinia ferrea*), *Solanum* sp., laburnums, jasmines, *Albizia* sp., *Tipuana* sp. (which is a declared weed).

They are able to buzz pollinate by grasping the flower with their legs and vigorously vibrating their thoracic muscles, thus agitating the flower via sound (sonication). Buzz pollination is important for wildflowers like *Melastoma affine* or flowers where the pollen is trapped in narrow tubes.

To attract females, the males emit a pheromone smelling of flowers and pollen. The males are seldom seen as they establish their territories high up in the crown of trees. They detect intruding males by the pheromones they emit and defend their territories.

Similar species of Carpenter bees in other countries are good pollinators of passion fruit.

14.22

14.23

14.24

PEACOCK CARPENTER BEE OR GREEN CARPENTER BEE — *XYLOCOPA (LESTIS)*

Xylocopa bombylans is found along the coast from Sydney to Cape York and *Xylocopa aeratus* is found around Sydney, on the Dividing Range up to approximately the Stanthorpe area.

The female of these spectacular bees is a glossy metallic blue with purple tints and the male a metallic green with yellow tints. Peacock Carpenter bees are 17 to 20 millimetres long.

14.22: Peacock Carpenter bee, *Xylocopa (Lestis) bombylans.*

14.23: Peacock Carpenter bee, female.

14.24: Peacock Carpenter bee, male in nest.

They cut 7 to 10-millimetre-wide nest burrows in the flower stalks of Grass trees (*Xanthorrhoea*) or in the soft pithy dead timber of *Banksia*, *Tristania*, *Acacia*, *Leptospermum* and *Casuarina*, where they make rounded cells for their eggs. The cells are stocked with pollen and nectar formed into bee bread before the female lays a large egg on top. She then closes the cell with a plug of chewed wood particles. The nest burrow may be shared by males and females. The emerging young bees are fed in the nest for some time before they venture outside.

Peacock Carpenter bees are able to buzz pollinate and are attracted to flowers needing buzz pollination such as *Hibbertia scandens* and *Solanum jasminoides*. Land clearing has caused the loss of these stunning bees from Victoria and mainland areas of South Australia although they can still be found on Kangaroo Island in South Australia.

14.25/14.26/14.27: Neon Cuckoo bee, *Thyreus nitidulus*.

CUCKOO BEE — *THYREUS*

The Cuckoo bees are stunningly beautiful but these bees lead a life of deceit! They behave like cuckoo birds, laying their eggs in the nests of other bees. The Neon Cuckoo bee, *Thyreus nitidulus*, with glittering metallic blue markings on a black body, lays its eggs in the nests of Blue-banded bees.

The Domino Cuckoo bee (*Thyreus lugubris*) with white patches on a black body, lays its eggs in the nests of Teddy Bear bees.

There is also the Chequered Cuckoo bee (*Thyreus caeruleopunctatus*) which has blue spots on a black body.

According to overseas studies the cuticle of Cuckoo bees is thicker than normal for defence when fighting host bees. Their larvae have huge, sharp mandibles to kill the egg/larvae of the host bee but sometimes they simply eat the provisions, resulting in a very stunted host bee emerging from the cell.

REED BEE — *EXONEURA*

There are 69 species of Reed bee recorded in the Australian Faunal Directory. Reed bees can be found in all states of Australia with most found in coastal areas of Victoria, New South Wales and Queensland.

Reed bees are slender bees measuring between 4 and 8 millimetres long. Most species have a red abdomen but some have a black abdomen. Many have white markings on their faces. Females have a sting but are not aggressive.

They nest inside dry, pithy twigs in plants such as raspberries and blackberries or in the dead fronds of tree ferns. Many nests can also be found in dead canes of the weed Lantana. They will also nest in small drilled holes in hardwood. They block the nest with their behind to stop ants or other predators from entering.

14.28/14.29: Reed bee, *Exoneura*.

The females do not make cells or store pollen or honey but feed the larvae individually as they lie in the nesting cavity.

Reed bees feed on both native and exotic flowers including *Alyssum, Acacia, Banksia, Eucalyptus, Grevillea, Leptospermum* and *Melaleuca*.

LEAFCUTTER BEE — *MEGACHILE (EUTRICHARAEA)*

There are believed to be about 27 species of Leafcutter bees from the genus *Megachile*. They are found in all states of Australia and can be found in both coastal and drier inland areas of Australia. The Leafcutter bees range in size from 6 to 15 millimetres. Most Leafcutter bees are black and have white or orange-gold stripes of hair on the abdomen. They carry pollen on special bristles underneath the abdomen. They can usually be distinguished from Resin bees by their relatively wide abdomen which tapers into a point, while resin bees have a more narrow and cylindrical abdomen. Leafcutters also alight on flowers with their wings spread, while Resin bees fold their wings.

Bee watchers often first discover these amazing bees when they notice rows of neat circular cuts on the edges of some leaves in their garden. Leafcutters use the discs of leaf as nest building material. They particularly like the soft leaves of roses, *Bauhinia*, *Desmodium* and *Buddleja*.

They use the cut leaves to make a tube as a nest for their eggs or line a hole in timber with it. Some species of Leafcutter bees nest in existing holes or crevices in timber or masonry. One species has been observed at the Australian Native Bee Research Centre (ANBRC) digging a shallow nest burrow in the soil of a pot plant.

14.30

14.30: Leafcutter bee, female, *Megachile*.

14.31: Leafcutter bee, female.

14.32: Leafcutter bee, female, *Megachile serricauda*.

14.31

14.32

14.33

Each nest is built by just one female. The leaves are cut in various shapes, round and elongated, to suit the construction of the cell for the egg. The cell is then stocked with a mixture of nectar and pollen in which the Leafcutter bee lays her egg. She cuts more circular leaves to close off the cell before constructing the next one until the hole or tube is filled. The hole is then plugged with rough leaf cuttings.

When the eggs hatch the tiny larva eats the provisions and when fully grown spins a silky cocoon and develops into a pupa then emerges as an adult bee. Immature bees may hibernate through winter and complete development into an adult the following spring.

A minute wasp from the genus *Melittobia* parasitises the Leafcutter bees and hundreds may be found in one cocoon eating the larva.

RESIN BEE — *MEGACHILE*

Resin bees are found in all states of Australia. They carry pollen on scopal hairs underneath the abdomen.

Resin bees come in many colours and sizes. For example there are large black 14-millimetre bees with white tufts of hair, and small 8-millimetre black bees with bright orange abdomens. Resin bees nest in pre-existing holes made by wood-boring beetles or moths, in gaps in timber or stonework and will use artificial nests.

They are called Resin bees because they collect resins and gums to construct their nest and build partitions between their brood cells and to seal their nest holes. They roll the resin with their front legs into a ball before taking it to the nest. Plant material mixed with resin may also be used to close off the nest.

The female Resin bee mixes pollen and honey into a paste called bee bread on which she lays her egg as food for the developing

14.33: Leafcutter bee leaving nest after it had fallen from its location.

14.34: Resin bee, *Megachile punctata*.

14.34

14.35: Resin bee, *Megachile punctata*. 14.36: Resin bee, *Megachile mystacea*, male.

larva. The eggs hatch into small white larvae, eat the food provided and then spin silken cocoons and develop into pupae before emerging as adult bees.

Resin bees hibernate during winter and emerge in spring. The nests can be parasitised by wasps.

Beekeepers sometimes notice Resin bees hanging around hives, trying to 'borrow' a little resin for their nests.

HOMALICTUS BEE — *HOMALICTUS*

There are about 40 species of *Homalictus* bees and they are found in all states of Australia, both along the coast and inland.

Although very small, ranging from 5 millimetres to 8 millimetres long, the glittering *Homalictus* bees come in a dazzling array of colours; 'golden blue', 'coppery red' and 'green tinged with purple, red or gold' are just a few of the colours listed by scientists. They carry pollen on unusually long, branched hairs under the abdomen.

Homalictus bees dig complex branching nests in the ground. Tiny oval brood cells are constructed in the nest shafts off the main shaft. Many females may live together in each nest, taking turns to guard the narrow nest entrance. One nest was found to be occupied by over 160 females!

Homalictus bees forage on a wide variety of flowers from many different genera such as mistletoe, *Angophora, Eucalyptus, Melaleuca, Tristania*.

The female lays a single egg onto a ball of prepared food, then seals the brood cell up and starts constructing the next cell.

Male *Homalictus* bees roost at night in the open, clustered together like other solitary native male bees.

14.37/14.38: Homalictus, *Homalictus urbanus*.

14.39: Halictid bee, *Halictidae*.

14.40: *Homalictus urbanus*.

14.41: *Lipotriches*.

14.42: *Lipotriches* sp.

14.43: *Lipotriches (Austronomia) phanerura.*

14.44: Female *Lipotriches* nest in burrows in the ground. At night, large numbers of males roost together in the open, clinging to grass or dry vegetation.

14.45: *Lipotriches* bee.

LIPOTRICHES — HALICTIDAE

Various species of *Lipotriches* bees occur in all states of Australia according to records on the Atlas of Living Australia website. They live in woodlands, forests, deserts and urban areas.

Lipotriches bees measure about 10 millimetres and carry pollen on scopal hairs on the hind legs. Some species of male *Lipotriches* have enlarged hind legs that may be used during courtship, according to overseas studies.

Most *Lipotriches* nest in burrows in the ground with each tunnel about 8 millimetres in width and widening a few centimetres below the surface to allow the guarding bee to retreat, allowing access to the entering bee. The main entrance is used by up to three females but each constructs her own side tunnel. The end of the tunnel contains a room where vertical brood cells are constructed. Pollen and nectar are deposited in each cell before an egg is laid. They guard the nest entrance, blocking it with their faces during the day and the tail at night. Some newly mated females spend the winter in their nest in semi hibernation.

At night large numbers of males cluster together on bare branches of trees, grass stems or other dry vegetation, the first arrivals attaching themselves by their mandibles and later arrivals clinging to the bodies of the first ones. Tarlton Rayment (*A Cluster of Bees*, 1935) wrote that some clusters may contain thousands of male bees.

MASKED BEE — *AMPHYLAEUS, HYLAEUS* AND *MEROGLOSSA*

Masked bees can be found throughout the world but the biggest variety is in Australia, with approximately 200 species making their home here. They can be found in every state and territory, from rainforest to deserts.

These slender black bees (most less than 10 millimetres long) are called 'masked bees' because they have pale markings on their faces. Many species also have a distinctive yellow spot on the thorax; some have a yellow collar.

Masked bees have very little hair and carry pollen to their nests by swallowing it. The nests are usually in deserted mud wasp nests, pre-existing holes in wood and pithy stems of dead tree fern fronds. The females excavate 4 to 9-millimetre-wide burrows in the stems, then line them with waterproof cellophane-like secretions, according to scientists Dr Katja Hogendoorn and Dr Allan Spessa. Brood cells are then constructed and provisioned with pollen and nectar before the egg is laid. The cell is then enclosed with further waterproof cellophane-like material before being sealed.

Adult Masked bees spend winter in their nest. In spring some bees will set up new nests and others will re-use the old nest. New nests are built by single bees while re-used nests are sometimes used by a pair.

14.46: *Lipotriches.*
14.47/14.48/14.49: Masked bee.

15.

Diseases of the honey bee

DISEASES OF BROOD

The key to the successful management of the health of a hive is for the beekeeper to inspect the colony and supporting frames closely during each hive opening and to look for the first signs of disease. Identified and treated early, it is often possible to minimise the effect of the disease or to eliminate it completely.

Minimising and treating bee disease within a colony needs to be the primary aim of every beekeeper, both amateur and professional. Without a healthy colony honey production will not be at its maximum, wasting both time and money for the beekeeper and, more importantly, the prospect of disease being passed on to other healthy colonies nearby is very real.

Many diseases have very subtle signs and new beekeepers in particular often have trouble recognising signs of a diseased hive. It is vital that the beginning beekeeper develop the habit of inspecting hives regularly so that experience is gained in looking at a healthy viable hive. Later on, if something changes within the hive and disease is suspected, it will then be much easier to recognising and identify any differences or changes, and seek advice from other beekeepers about the possible causes.

An important aspect of minimising the transmission of disease by beekeepers is to keep tools and equipment clean. If you suspect a hive that you have inspected is infected with American Foul Brood (AFB) or European Foul Brood (EFB), once a hive tool has been used inside the hive it should be sterilised before use on another hive. Sterilisation can be performed quickly by placing the exposed end of the hive tool inside the hot smoker when it is being

15.2

15.1: Diseased drone cells.

15.2: Damage caused by Greater Wax Moth.

carried or, alternatively, using a strong disinfectant such as household bleach in a container in which the hive tool or multiple tools can be kept between hive inspections.

GOOD MANAGEMENT PRACTICES TO MINIMISE DISEASE IN THE COLONY

Another aspect of keeping colonies free of both brood and adult disease is good management practices by the beekeeper. These include:

- ensuring appropriate nutrition
- making sure that queen bees are young
- keeping strong hives, possibly by merging with another clean colony if required
- minimising dampness and condensation in the hive, particularly in colder areas during winter
- practising comb rotation by replacing two or more old brood frames every year with new frames and foundation
- trying not to move frames of honey or brood between hives since this may aid in the spread of disease
- minimising stress on the colony by not unduly moving the hive or opening the hive too often.

AMERICAN FOUL BROOD (AFB)

American Foul Brood, or AFB, caused by the bacterium *Paenibacillus larvae*, has become one of the most serious bee diseases in Australia. AFB can cause extensive losses in both amateur and commercial apiaries although severe outbreaks can be minimised by good management practices, regular checks for the disease, and destroying or irradiating hives when the disease is confirmed.

American Foul Brood spores can remain dormant for over 50 years in a vegetative state on old equipment that was once used with infected colonies. Once larvae are more than three days old they are far less susceptible to infection by AFB. American Foul Brood is a notifiable disease and your state or territory's apiary inspectors need to be informed if you suspect an outbreak in any of your colonies.

SIGNS OF INFECTION

During the spring and autumn a thorough examination of brood frames for signs of the disease will need to be conducted. It should be noted that AFB can occur in hives at any time of the year. A close inspection of the frames may detect early infections, which often appear in only one or two larvae or pupae.

To the beginner beekeeper, American Foul Brood (AFB) and European Foul Brood (EFB) may appear similar and are often confused with each other. The following points may help to differentiate between them.

- In heavily infected colonies the brood has a scattered, uneven pattern due to the intermingling of healthy cells with diseased cells.
- Brood infected with AFB generally die after the cell has been capped. This is different to EFB, where brood usually die before the cell is capped.
- AFB brood frames may show capped cells of dead brood with punctured and sunken cappings.
- This peppered appearance of the brood usually allows AFB to be distinguished from EFB. In AFB the cappings are discoloured, while in EFB the cappings are not normally discoloured to any great extent.
- Both AFB and EFB can result in brood under sealed cells exhibiting a sunken dark appearance with perforated cappings.
- Brood affected with AFB will change colour from a healthy pearly white to a darker brown as the disease progresses.
- Brown dead brood probed with a matchstick may show signs of a ropy consistency in infections of both AFB and EFB. If a cell infected with AFB is tested with the matchstick test the resulting ropiness may be 3 to 5 centimetres long, while the ropiness test on EFB affected brood may result in little or no stringing out of the cell contents.

- Colonies infected with AFB may give off a putrid fish-like odour while colonies infected with EFB often give off a slightly sour or sometimes rotten faecal odour.
- After about a month brood that have died from AFB dry out to a dark scale which adheres to the wall of the cell. Brood that have been killed by AFB lie on the bottom of the cell whereas those that have been killed by EFB may lie in a more unnatural sideways position as if they had been wriggling out of their normal position at the time of death.
- The tongues of brood killed by AFB are often sticking out and sometimes attached to the roof of the cell. This is not the case with brood that have been killed by EFB.
- The only accurate diagnosis is by laboratory analysis. There are, however, easy-to-use field diagnostic kits made by VITA available at a small price from beekeeping suppliers that provide reasonably accurate testing for AFB.

15.3: The cappings on cells containing American Foul Brood (AFB) infected pupae are typically sunken and discoloured. Larvae infected with AFB usually die after the cell is capped although some strains are now appearing where the brood dies before the cell is capped.

15.4: The remains of pupae killed by American Foul Brood (AFB) are ropey and can be pulled out using a matchstick, the aptly named matchstick test for AFB.

15.5: Pupae that have recently died of AFB have a characteristic false tongue that attaches itself to the roof of the cell.

15.6: Inexpensive kits can now be purchased to test for both American Foul Brood (AFB) and European Foul Brood (EFB).

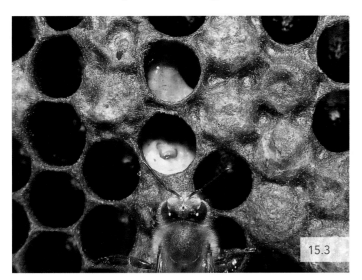
15.3

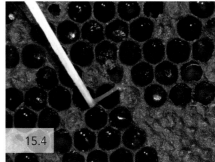
15.4

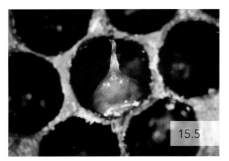
15.5

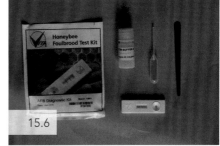
15.6

SPREAD

The control of AFB is very much in the hands of the beekeeper as the most common method of spread is via the use of contaminated equipment in healthy colonies. The use of infected hive tools, gloves or the interchange of hive parts within the apiary should be done with the greatest of care to avoid spreading any brood disease, not just AFB. Colonies may also become infected if bees are fed contaminated honey, rob honey from contaminated hives or feral colonies, or if infected comb is left on the ground within the apiary for other bees to feed off. Young larvae less than 53 hours after egg hatch are the most susceptible to AFB. Careful inspection and good hive and apiary management are the keys to preventing serious outbreaks of this disease. Transmission of AFB between feral colonies is rare and the disease mainly affects colonies belonging to professional and hobby beekeepers.

TREATMENT

First, the infected hive needs to be sealed and the colony killed with an insecticide to stop infected bees from escaping or the infected honey from being robbed. Contaminated materials such as the hive and frames are destroyed by burning. Burning of the hive needs to be carried out in a pit to contain any wax and honey left over. The remaining ashes are covered with 30 cm of soil to stop robber bees from stealing any residual honey and spreading the disease.

Sterilisation of contaminated hives using gamma radiation is also used to kill bacteria, although this option is only available in some of the larger capital cities in Australia. Before infected hives are sterilised, bees and honey must be removed and destroyed using the above-mentioned techniques. Frames, boxes, hive covers, bottom boards and queen excluders are prepared for irradiation by wrapping them in a double layer of thick plastic garbage bags. After sterilisation, hives can be restocked with disease-free bees.

Both of these methods successfully minimize the incidence of the disease and are the only techniques approved in each state and territory.

In Tasmania, treating mildly infected colonies with the antibiotic Oxytetracycline hydrochloride (OTC) is legal to control AFB. As in all other states and territories of Australia, severely infected colonies and hives must be destroyed.

The announcement of a vaccine fed to queens in the bee candy included in every queen cage is an exciting step in managing AFB. Vaccination is still at the trial stage, and beekeepers will watch with interest to see if vaccination becomes an effective control method for this and other diseases.

EUROPEAN FOUL BROOD (EFB)

European Foul Brood, or EFB, is caused by the bacterium *Melissococcus pluton* and has become a serious bee disease in Eastern Australia. European Foul Brood will often cause extensive losses in both amateur and commercial apiaries and again outbreaks can be minimised by regular disease checks, early detection, and good management practices by the beekeeper.

EFB can remain dormant for over three years in a vegetative state on old equipment that once was used with infected colonies; it is highly contagious with all stages of larval development susceptible to infection. The disease is found throughout Australia.

EFB is a notifiable disease and your state or territory's apiary inspectors need to be informed if you suspect an outbreak in any of your colonies.

SIGNS OF INFECTION

Fortunately, EFB diseased colonies are usually easily recognised as being diseased although identification of the cause of the disease may be more difficult for the amateur due to its similarity to American Foul Brood. The signs associated with infection by EFB are very similar to those associated with AFB. As discussed above, the two main differences are:

- Brood infected with AFB generally die after the cell has been capped. This is different to EFB where brood generally die before the cell is capped.
- Brown dead brood probed with a matchstick may show signs of a ropy consistency in infections of both AFB and EFB. If a cell infected with AFB is tested with the matchstick test the resulting ropiness may be 3 to 5 centimetres long, while the ropiness test on EFB may only result in little or no stringing out of the cell contents.

Keep in mind that diagnosis of EFB based solely on the signs described above is not always reliable. The only accurate

15.7: Larva infected with European Foul Brood (EFB) usually die before the cell is capped.

15.8: Larvae infected with European Foul Brood (EFB).

15.9: European Foul Brood scale with the characteristic gondola shape.

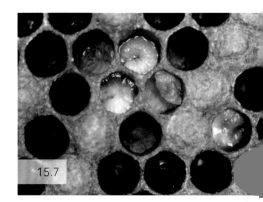

15.7

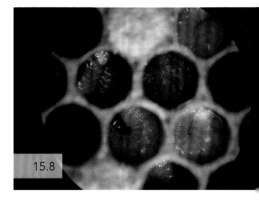

15.8

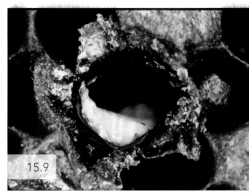

15.9

diagnosis for EFB is by laboratory analysis, which beekeepers can access by submitting a comb sample or a smear of brood cell contents on a glass slide to a diagnostic laboratory. An EFB self-diagnostic kit is sold by VITA, and there is a similar kit for AFB diagnosis.

SPREAD

European Foul Brood is highly contagious and the causes of spread are the same as for AFB. Infected colonies can survive for long periods with low spore counts of EFB without showing any serious signs of the disease. Sudden outbreaks of the disease can then occur due to a build-up of spore numbers; usually these outbreaks are caused by colony stress resulting from changed seasonal conditions, poor nutrition and moving bees. Young larvae less than 48 hours old are the most susceptible to EFB.

European Foul Brood enters the gut of the larvae and essentially competes with it, consuming food that is needed by the young brood. As a result, developing brood are left with insufficient food on which to survive and literally starve to death. Some experienced professional beekeepers say that EFB is likely to show its presence in a hive when there is a shortage of pollen coming in. If there is an abundance of pollen and large quantities of royal jelly are being fed to the larvae, there may be sufficient food for both the EFB and the larvae so that the disease will not show itself by harming the brood.

TREATMENT

The only antibiotic recommended for the treatment of EFB is oxytetracycline hydrochloride (OTC), sometimes called Terramycin. Oxytetracycline hydrochloride is available only on prescription from a veterinarian or with an Order to Supply from an apiary officer of a state or territory government. Beekeepers have been known to treat infected colonies with too much OTC resulting in honey containing high residues of the antibiotic.

While there may be times when antibiotic treatment would appear to be the only answer, its use is becoming increasingly less attractive due both to concerns that the honey may become contaminated and to the possibility that overuse will result in the development of antibiotic resistant strains of EFB bacteria. For the hobby beekeeper, the use of antibiotics has never really been a practical solution mainly due to the difficulty of obtaining the appropriate prescription. Alternative methods of treatment available to the hobbyist are described on p. 249.

If ample honey is stored in the hive a shortage of nectar should not be a problem, but good-quality pollen is another matter. Pollen is available either when it has been stored by the bees or when it is available from flowering plants. A good supply of pollen is essential in order to provide adequate protein levels that result in a well-balanced group of amino acids, important to reduce any nutritional imbalance that will stress the bees. A temporary lack of pollen can be overcome by feeding the bees either previously collected pollen or a pollen substitute.

SACBROOD VIRUS (SBV)

Sacbrood, or SBV, is not common in Australia. The infection is caused by a virus and infected larvae typically die shortly after capping but before the change to the pupal stage of their development in the capped cell.

Sacbrood infected larvae form a plastic-like outer layer or skin that contains a watery fluid, giving the infected larvae a characteristic watery bloated look. Sacbrood infections are most common during the early spring. The colony at this time is very susceptible to both chill and nutritional stress, either of which can lead to further progression of the Sacbrood disease.

SIGNS OF INFECTION

Sacbrood mainly affects female worker larvae although it may sometimes occur in male drones. The disease affects uncapped pupae as well as sealed brood that are seven to ten days old. Brood comb contaminated with Sacbrood show some degree of irregularity of brood pattern similar to EFB or AFB. Dead brood will be found scattered among healthy brood in the comb and their cappings may be discoloured, sunken, perforated or removed by the bees. Brood almost always die after the cell has been capped and generally not during the uncapped larval stage.

Dried brood that have died of SBV are located on the bottom of the cell, become brittle, and retain the upturned 'gondola' shape of the body. The dried scale can easily be removed without damaging the cell wall. Infected brood are typically odourless, though in advanced stages an occasional putrid smell may be detected.

SPREAD

It is thought that nursery bees probably become infected with Sacbrood when cleaning out the cells of infected pupae and then pass the virus on to non-infected larvae when they feed larvae with brood food such as royal jelly from their hypopharyngeal glands.

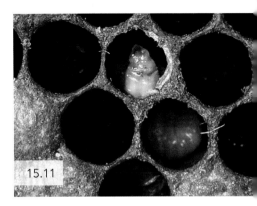

15.10

15.11

15.10: Sacbrood virus.

15.11: Although this larva looks as if it is infected with Sacbrood virus, the colony only tested positive for infection with European Foul Brood (EFB).

The virus breeds rapidly, leading to death shortly after capping. The virus may remain active for up to four weeks in the larval remains or in honey or pollen.

TREATMENT

The incidence of SBV in most colonies is low because adult bees normally detect and quickly remove infected larvae. Good management practices by the beekeeper as detailed on p. 249 can help to alleviate Sacbrood.

CHALKBROOD

Although commonly found in hives, Chalkbrood is not usually a serious disease of honey bees and only very rarely kills an entire colony. If, however, the bees have a predisposition to the disease it can cause a gradual deterioration of the colony by killing a larger number of brood, thereby affecting adult bee numbers and resulting in a loss of honey production. Chalkbrood is caused by the fungus *Ascosphaera apis* and there are signs that its incidence is increasing around the world, probably due in part to migratory beekeeping and to poor hygiene in the practices of some beekeepers. There has also been the suggestion that the importation of infected pollen and foundation from abroad may be an additional cause of the increase in incidence of the disease. The fungus can remain active on hives and other surfaces for up to fifteen years.

SIGNS OF INFECTION

Chalkbrood can infect both unsealed larvae as well as sealed pupae. The main signs of the disease are as follows.

- Infected brood are called mummies. When mummies are removed from their cell they appear to be solid, similar to lumps of chalk.
- Partially chewed down mummies are often seen in open cells.
- Mummies vary in colour from white to dark grey or black.
- Many of the sealed cell caps may have a small pinprick-sized hole.
- Dead brood will be dropped onto the hive floor by nursery bees and later moved outside the hive entrance by worker bees, leaving a litter of chalk-like mummies on the ground.
- A hive heavily infected with Chalkbrood will show a patchy scattered brood pattern.
- Chalkbrood is more prevalent during the spring since fungal growth is enhanced during cool, damp conditions, particularly in poorly ventilated hives.
- Chalkbrood can be spread by workers, drones or queens.

15.13

15.14

15.12: One of the signs of Chalkbrood infection is hard, off-white, chalk-like cadavers at the hive entrance.

15.13: Pupae killed by Chalkbrood removed from cells.

15.14: The tell-tale sign of Chalkbrood: white, hard cadavers peppered through the brood cells.

SPREAD

Chalkbrood fungus, *Ascosphaera apis*, can initially be passed to an uninfected colony in a number of ways, namely by forager bees returning with infected pollen or nectar, robber bees stealing honey from an infected colony or by the drift of bees from an infected colony to an uninfected colony. This latter scenario may arise when hives are moved with an open entrance or by the beekeeper not following hygienic management practices. The fungus is passed orally to uninfected bees through contaminated food or by worker bees cleaning the inside of an infected hive causing the fungus to be consumed. While adult bees are not susceptible to Chalkbrood disease they are the carriers and pass the fungus on to larvae that consume it, become infected and eventually die.

Young larvae between the ages of one to four days are the most susceptible to Chalkbrood disease. Once the Chalkbrood fungus has been consumed by the larvae and is inside its mid-gut, the fungus penetrates the gut wall and grows inside the body cavity of the larvae. As the infection develops the fungus eventually breaks out of the body cavity, penetrates the outer surface or skin, initially at the posterior or anal end but eventually breaking out over the body, covering the brood with a thick white layer of mycelium, similar to household mould on fruit and bread. At the end of each string of mould on the outside of a blackened corpse there will be a new infectious spore, with as many as 100 million to 1 billion active spores on the outside of each black corpse ready to be ingested by unsuspecting adult bees.

Honey bee brood that have died with Chalkbrood infection are initially swollen in size, taking up the entire volume in the cell. Eventually the corpse dries out leaving the characteristic chalk-like pellet normally associated with the disease.

TREATMENT

There are no chemical treatments for Chalkbrood although feeding sugar syrup to infected colonies has been shown to be beneficial in helping bees to fight this disease. Like so many other honey bee diseases, good management practices by the beekeeper lie at the frontline of control and treatment. These management practices are detailed on p. 249.

ADULT BEE DISEASES

NOSEMA

Nosema is a fungus that infects the gut of the adult honey bee. Although it is rarely fatal for the colony, Nosemosis, as infection with *Nosema* is called, can cause the colony harm, resulting in a significant loss in honey production and compromised bee health including the death of bees, particularly during the spring. *Nosema* comes in two strains:
- *Nosema apis*
- *Nosema ceranae.*

Both varieties of *Nosema* are probably present at low levels in all honey bee colonies but rarely develop to become a significant infection unless conditions favour development of the disease.

Historically, it was *Nosema apis* that infected honey bees but more recently *Nosema ceranae* has jumped species from the Asian honey bee to the European honey bee. The more recent *Nosema ceranae* fungal infection is the more lethal of the two and is believed to be causing Australian beekeepers the majority of the problems that are associated with *Nosema*. Infection with *Nosema* is difficult for the beekeeper to diagnose and the only accurate test is to send a number of infected or suspect bees to a laboratory for analysis.

Generally *Nosema apis* is more of a problem in regions with longer winters if bees are unable to leave the hive to defecate due to poor weather. *Nosema ceranae* appears to be more prevalent in warmer climates.

Although *Nosema apis* has been known to infect honey bees for many years, the recently arrived *Nosema ceranae* is proving to be far more harmful to bees and has the added ability to displace *Nosema apis* within colonies. I will describe *Nosema ceranae* as it is the most prevalent strain with the most impact on Australian bee populations.

SIGNS OF INFECTION

There are no classic outward signs of Nosemosis and an infection may go undetected until the disease progresses sufficiently for the colony to be significantly affected. For this reason infection with *Nosema* is often called the silent killer. The main signs to look for are:

- a declining population of bees, particularly during the winter
- bees that look shiny
- a reduction in the egg-laying ability of the queen and her possible supersedure
- general lethargy within the hive.

Unlike *Nosema apis*, *Nosema ceranae* typically does not cause diarrhoea and infection is more difficult for the beekeeper to detect and diagnose. Infection with *Nosema* is more prevalent when nutrition is poor and weather conditions are wet and cold.

TREATMENT

In order to minimise the effect of *Nosema* on the colony, good management practices should be used as described on p. 249.

Nosema build-up is particularly strong during winter months when the bees are restricted from leaving the hive to defecate due to cold and poor weather. It is thus important for the beekeeper to ensure that hives are placed in warm, open areas with the entrance facing the morning sun. Also, raising the hive off the ground to enable good air circulation helps the bees fight off the disease. Many experienced apiarists believe that the best way to treat infection with *Nosema* is to leave the hive alone for a few months while ensuring that the conditions described above are met.

Hives that have been contaminated by *Nosema* may be thoroughly cleaned with glacial acetic acid. Spores of *Nosema ceranae* are susceptible to death by freezing or by refrigeration for a week.

The more a colony with *Nosema* is worked to produce honey or for pollination the worse the infection is likely to become.

SPREAD

Nosema spores are spread by worker bees eating infected faeces; the faeces may be ingested when cleaning the hive (but not by cleaning brood cells since brood are rarely infected), or by coming into contact with bee faeces on flowers when collecting food or at water sources. Once inside the mid-gut the *Nosema* spores infect the cells lining the stomach and multiply rapidly. Within six to ten days the infected cells become filled with new spores, the cell dies and bursts, with each bee releasing 30 to 50 million new

15.15: A bee infected with Deformed Wing Virus showing the classic signs of deformed wings and a misshapen body.

15.16: Testing of bees from the colony that hosted this bee apparently infected with Deformed Wing Virus (DWV) did not show any signs of the virus. Similar signs to infection with Deformed Wing Virus can often have other causes such as brood chilled during cold weather.

spores in their faeces ready to infect other bees. Infection with *Nosema* generally affects worker bees, probably because neither drones nor queens take part in the cleaning activities within the hive or foraging outside of it. Newly emerged bees are seldom infected by *Nosema*, since they have not had the opportunity to come in contact with, and digest, faecal matter. Infection with *Nosema* severely restricts the ability of bees to digest pollen, thus limiting their own growth and limiting their ability to make royal jelly and other foods. These foods are essential for feeding to:

- developing larvae
- the queen to support egg production
- drones to ensure maturation of sperm
- other adult workers.

DEFORMED WING VIRUS (DWV)

Although surveillance for Deformed Wing Virus, DWV, is ongoing, field studies by CSIRO during 2014 failed to detect Deformed Wing Virus in Australia. In spite of this, adult bees are sometimes observed in Australian hives that show the classic symptoms of Deformed Wing Virus — deformed, stubby wings and/or misshapen bodies. These apparent signs of DWV are more likely to be the result of genetic birth defects or the larvae being too cold inside the brood cell.

Deformed Wing Virus, though, is one of a few bee viruses easily recognisable due to its well-defined signs in infected bees. Typical signs of Deformed Wing Virus infection include:

- stubby, useless wings
- shortened, rounded abdomens
- discolouring of the adult bee's body
- paralysis.

Deformed Wing Virus is known to be transferred by the *Varroa destructor* mite. Studies of *Varroa destructor* have showed that up to 100 per cent of the mites are carrying the virus. Since infestation with *Varroa* mites is one of the main ways in which Deformed Wing Virus is passed to bees, ongoing monitoring for this virus is regarded as being a critical part of surveillance for any *Varroa* incursion into Australia.

SUMMARY

Many of the diseases that currently infect honey bees are assisted in their spread by the beekeeper. As a result, good management practices are at the heart of any disease minimisation work.

The most common diseases of honey bees are:
- American Foul Brood (AFB) — a bacterium with resistant, long-lived spores that kills capped brood
- European Foul Brood (EFB) — a non-spore-producing bacterium that generally kills uncapped brood
- Sacbrood — a virus that kills brood
- Chalkbrood — a fungus that kills brood
- *Nosema* — a fungus that affects adult bees. Present in Australia in two forms: *Nosema apis* and *Nosema ceranae*
- Deformed Wing Virus (DWV) — a virus that causes adult bees to emerge with deformed wings and bodies.

All the above diseases except Deformed Wing Virus are found in Australia. American Foul Brood is the only disease for which the affected colony will need to be destroyed to minimise its spread.

16.1

16.2

Parasites of the honey bee

Once *Varroa* had been detected in honey bee colonies in Europe and North America there was a concerted effort globally to identify all of the mites that live with honey bees. To date, about 100 mites have been identified although only three (not including the pest, the Small Hive Beetle) have been found to cause economic loss to beekeepers. The three mites that cause economic loss are *Varroa destructor*, Tracheal Mite (*Acarapis woodi*) and the *Tropilaelaps* mite. Both *Varroa destructor* and to a lesser extent Tracheal Mites are causing severe economic loss to beekeepers across the world. *Tropilaelaps* is restricted to Asia and is not yet a problem for beekeepers in Europe, North America, Australia or New Zealand. *Tropilaelaps* has wiped out honey bees in many parts of New Guinea and is severely limiting European honey bee beekeeping in Asia.

Both *Varroa* and *Tropilaelaps* are discussed below although Tracheal Mite, the least destructive of the three, is not discussed here. The reason for this is that the effect of the Tracheal Mite on honey bees is much less severe than either *Varroa* or *Tropilaelaps*. Also Tracheal Mite is not as yet found in either Australia or New Zealand.

VARROA

There are two species of *Varroa* that infect the European honey bee. These are:

- *Varroa destructor*
- *Varroa jacobsoni.*

16.3

16.1: A *Varroa* mite on the back of a worker bee.

16.2: A *Varroa* mite on the abdomen of a worker bee.

16.3: *Varroa* mites infesting a colony.

Varroa destructor was discovered in Australia, near Newcastle, NSW, in 2022. Government officials and beekeepers have made a significant effort to eradicate the incursion, although it remains to be seen whether the mite has been eradicated from the feral population. If not, the mite will spread slowly at first, then more quickly due to migratory beekeeping, throughout eastern Australia. There exist strict quarantine regulations about the movement of hives from eastern Australia to Western Australia, so Western Australia is likely to remain *Varroa*-free for a number of years. *Tropilaelaps* has still not been found in Australia.

16.4

Varroa destructor has long been a destructive parasite of the European honey bee, having jumped from its original host species, the Asian honey bee, *Apis cerana*, probably in the 1940s but perhaps much earlier. More recently, *Varroa jacobsoni* has been found to infest European honey bees in Papua New Guinea and this development is causing concern to beekeepers as the new parasitic mite adds additional strain to an already stressed species.

SIGNS OF INFESTATION

Varroa destructor is large when compared to other mites and can easily be seen with the unaided eye. The width of the female mite is larger than her length, about 1.6 millimetres x 1.1 millimetres, with the female having a reddish colour to her body. Adult females of *Varroa destructor* move into brood cells in order to reproduce, and newly emerged female adults may also be seen walking rapidly on

16.5

16.4: *Varroa* mites on a larva.

16.5: Close up of *Varroa* mites photographed from the top and underneath.

the surface of brood comb before entering another brood cell to reproduce. Individual mites are often seen attached to adult bees, clinging mostly to the bees' abdomen where they feed on haemolymph (blood).

When a mature female *Varroa* mite enters a brood cell she first lays a male mite egg and then about six female mite eggs. If the initial female mite is the only *Varroa* to have entered the brood cell before it is capped, the male mite that she first lays is capable of fertilising his mother's daughter mites and expanding the number of viable *Varroa* mites in the colony. If other female *Varroa* mites enter the brood cell before it is capped the male mites can fertilise the daughter mites of the other *Varroa* mothers and bring greater genetic diversity to the reproduction process.

Adult *Varroa* males and *Varroa* male and female nymph are not seen outside of a brood cell as they are unable to survive without the wet and humid conditions found inside the capped cell. Immature mites are white in colour while adult females are reddish brown; adult males are smaller than females and, as previously noted, are not visible as their entire lifecycle is spent inside brood cells.

A female *Varroa* mite in a capped worker brood cell on average produces 1.4 to 1.5 mature daughter mites that emerge from the cell. The daughters remain on nurse bees for up to nine days before entering another brood cell to lay further *Varroa* eggs. Honey bee worker brood emerge from their cells after about twelve days as pupae while drones emerge after about fifteen days from their pupal cells. Since a fertile *Varroa* mother will lay several eggs inside a capped brood cell, the longer gestation period for male drones to emerge as adults means that about 2 to 2.5 mature *Varroa* daughters will emerge from a male drone cell with the drone. It is the number of capped drone cells inside a colony that largely determines how many mature *Varroa* adults will emerge and infect the hive.

The *Varroa* mite attacks young brood in a capped cell by piercing the soft skin of the brood in order to feed on fat bodies. Damage to adult bees is usually restricted to heavy infestations when a substantial proportion of the next generation is parasitised. If only a single mite enters the brood cell, damage to the bee may not be visible although the adult bee will have a much shorter lifespan than unparasitised bees. If multiple *Varroa* mites enter the brood cell, the brood larvae may die and any adult that emerges may be deformed with a shortened abdomen or deformed wings. Deformed wings are usually an indication of infection with Deformed Wing Virus but could also be caused by other factors such as chilled brood. In addition to physical signs, the behaviour of the adult bee may also be affected; for example, during orientation in flight or by experiencing difficulty in gathering food.

A *Varroa* infestation is a more complex parasitism than just the mites sucking brood fat bodies. During feeding the mite also injects viruses such as the Israeli Acute Paralysis

16.6

16.7

21

16.6: Sugar Shake Test 1. A cup full of bees is placed inside a container with a sieve built into the lid. Icing sugar is added to the bees and the container is gently rolled so that all of the bees become coated in icing sugar.

16.7: Sugar Shake Test 2. Next the container is turned upside down over a bowl of water and gently shaken. Any mites will become dislodged from the bees and will fall into the water, leaving the bees still inside the container. The mites, if present, will float on the surface of the water and can easily be detected.

Virus (IAPV) or Deformed Wing Virus (DWV) into parasitised brood. These and other viruses often have a significant impact on the bee's health, causing paralysis in adults, deformed wings, as well as severely reducing the lifespan of the infected bee.

There are several ways to monitor for *Varroa* although the most reliable but time-consuming method is to randomly open capped brood cells, particularly drone brood cells, and look for *Varroa* mites. The brood needs to be removed from the cell with a forceps and the brood and cell inspected for the mite. About 100 to 200 cells must be inspected in this way before an accurate assessment can be made of the scale of any *Varroa* mite infestation.

An alternative method is the Sugar Shake test. Three hundred or so bees (½ a cup of bees) are poured into a jar or container with a metal gauze lid that has a sufficiently fine mesh to enable the mites, but not the bees, to fall through the mesh. Once inside the jar the bees are covered with icing sugar and shaken to dislodge the mites. Then the jar is inverted so that the mites fall through the mesh into a bowl of water or onto a sheet of white paper. The beekeeper then inspects either the water surface or the sheet of paper for any mites that have fallen through.

16.8

Another method is to construct a tray that fits into the base of the hive. The tray is covered with mesh about 2 millimetres wide held about a centimetre above the floor of the tray. The tray is made of white or light material so that any *Varroa* mites falling onto it can easily be seen by the beekeeper. Once the tray is placed inside the base, it should be inspected after one to three days for any signs of infestation. Often either a sticky board or gooey substance is placed below the mesh to trap fallen mites.

The control of *Varroa destructor* is one of the most demanding and time-consuming tasks faced by beekeepers in countries where this mite is present. Not only do they have to manage the infestation but the learning curve to master the necessary skills to keep the mite under control can be quite a steep one.

SPREAD

Varroa was discovered in Auckland, New Zealand, in April 2000. A subsequent survey found that the mite had already spread to large areas of the North Island and that eradication was not feasible. To slow the spread of the mite further south a line was

16.8: *Varroa* mites that have fallen to the bottom of a hive and become stuck on a sticky board on the base.

16.9: A vented base board used in the detection and control of *Varroa*. The mesh is too small to allow bees to pass but *Varroa* mites can easily drop through. Underneath the mesh can be placed a sticky board or the *Varroa* can be allowed to fall through to the ground outside of the hive where they will die.

16.10: Vented bases used to help control Varroa also provide ventilation to the hive during hot summer days. The different colour markers on this hive facilitates use of the base to hold three separate queen rearing nuclei, each with its own entrance marked with a different coloured design.

drawn across the centre of the North Island across which it was not permitted to move hives. The line stayed in place from April 2000 to September 2003. At that time there was estimated to be 10,000 infested apiaries north of the line and 100 infested apiaries south of the line. It was decided that since there were already infested apiaries south of the line there was little point in keeping the quarantine zone. It has been estimated that the use of quarantine delayed the spread of *Varroa* by about two years.

It was initially believed that the water barrier of Cook Strait that separates the North and South Islands of New Zealand would form a natural barrier to the spread of the mite. Even though Cook Strait is 22 kilometres wide at its narrowest point, *Varroa* was found in Nelson at the top of the South Island in 2006. In 2008 all restrictions on moving bees and associated products were abandoned and the mite can now be found throughout much of the South and North Islands of the country.

In the United States, commercial movement of infested colonies for migratory beekeeping spread *Varroa* throughout the country in five years. At the local level, the major cause of *Varroa* spread is by other bees robbing collapsed colonies in search of honey. Other causes of the spread of the mite are by infested bees drifting to other hives or by foraging bees leaving mites on visited flowers, where they are picked up by other foragers. An infested queen used during re-queening or the merging together of two hives can also spread the mite. Beekeepers can help control the spread of *Varroa* by ensuring that sound management principles are used that minimise its spread.

In the early stages of an Australian incursion the beekeeper will look to the government apiary inspector, beekeeping associations and beekeeping supply stores for advice and management information. The experiences of other countries in the control and management of *Varroa* will become part of our own critical learning curve.

Judging by overseas experience, once *Varroa* reaches Australia and gains even a small foothold so that it cannot be eradicated, the spread of the mite throughout the country will be rapid. Experience in other *Varroa*-affected countries also indicates that if a feral colony is infested with *Varroa* it will only live about two to three years before the mite causes sufficient weakening for the colony to collapse and die. On the positive side, even though infestation by *Varroa* is a serious and time-consuming problem, beekeepers with hives affected by the mite often report increased honey production with a smaller number of hives. The reasons given for this increased production vary from the ability and management skills of the individual beekeeper to other factors such as fewer feral colonies competing for nectar and pollen. The extra effort and time required for effective management control in each hive no doubt also results in beekeepers often reducing hive numbers to deal with the added workload.

TREATMENT

Where V*arroa* has been introduced into other countries, within a few years survival of the resident European honey bees has only been achieved through ongoing management by the beekeeper. Most colonies of bees have required significant and expensive levels of management to survive while feral colonies died out within two to three years. A small number of feral colonies, however, have now returned to the United States and other countries where the mite is endemic. These colonies are still infested with *Varroa* but the mites are not as harmful to the bees as they once were. In some cases the mite became less harmful to the colony; in other cases the colony became more resistant to infestation. Feral bees that have developed resistance to *Varroa* are commonly called 'survivor stock'.

Since *Varroa* can be spread by the activities of the beekeeper, including the practice of migratory beekeeping, beekeepers will need to review the way that they manage their colonies, as this is one of the key tools in minimising the spread and the impact of the mite.

Other management techniques that are being used overseas to minimise the growth of the mite in hives is to disrupt their breeding patterns by inserting drone brood comb, as it is the drone cells which the *Varroa* mite prefers for reproduction. The drone comb is then removed, destroying the drone brood cells before the mite-infested drones are able to hatch out.

Miticides are still the most popular way to control *Varroa* overseas. The term miticide refers to any chemical that is used against mites such as Tracheal Mite, *Varroa* or *Tropilaelaps*. The term insecticide is used to describe a chemical that is used to control insects such as the Small Hive Beetle. Many proprietary chemicals, such as Apithor, may be used as both a miticide and an insecticide.

Although the use of chemicals to control *Varroa* risks contaminating honey and wax with lethal pesticides, if used correctly by carefully following the instructions on the packet these risks can be minimised. It should be noted that medicating the hive should not be done unless it is absolutely essential. Applying chemicals to a hive as a protective measure without any signs of *Varroa* is counterproductive since there will be toxic residues left in the hive.

A further major problem with applying chemicals to control mites is that frequent applications will result in the mites becoming resistant to the chemical over a relatively short time. In order to minimise the possibility of *Varroa* or other mites developing resistance to a particular chemical treatment the miticides used need to be rotated with the use of around three differently composed chemical treatments in order to delay the build-up of resistance to a single treatment. Treatment is usually effected around two to three times a year so the possibility of resistance build-up is an ongoing issue.

The chemicals used to treat *Varroa* vary in their toxicity to both bees and to humans. The naturally occurring organic chemicals oxalic acid and formic acid are relatively safe to bees at lower concentrations but not to beekeepers and need to be handled with care. Chemicals such as Apiguard, Apistan and Apivar should not be used in hives if there are supers containing honey present. The resulting honey may become chemically contaminated and unfit for human consumption.

Varroa was detected in New South Wales in 2022. The NSW incursion reminds all beekeepers that constant vigilance is needed when monitoring pests and diseases in their colonies. If this incursion is not eradicated, there will probably be further incursions that will eventually result in the mite becoming endemic. In addition to *Varroa*, many other pathogens, such as *Tropilaelaps* and Tracheal Mite, pose a threat to Australian beekeeping, and beekeepers and quarantine inspectors must remain alert to the possible presence of these parasites.

Apart from chemical miticides, many beekeepers in the United States and Europe are now working to develop *Varroa*-resistant or tolerant lines of bees. This has proven to be an effective, although currently underused, method of keeping *Varroa* and other diseases in check and I expect to see much progress in this area in the years ahead. Using a genetically tolerant strain of queen has the major added advantage of keeping lethal chemicals away from the hive as the chemical solutions applied to *Varroa* have now become part of the ever-increasing problem of hive contamination.

When the mite is detected on our shores a properly formulated and managed integrated pest management plan (IPM) will need to be in place to help manage the incursion and spread. There is no doubt that the mite's arrival will kill tens of thousands of colonies and decimate the feral bee population. Overseas experience indicates that many commercial beekeepers' livelihoods will be severely affected, at least during the early years of upskilling required to control this parasite.

TROPILAELAPS MITES — THE ASIAN MITE

Tropilaelaps mites are similar to *Varroa* and are parasites of honey bee brood. *Tropilaelaps* use the giant Asian honey bee, *Apis dorstata*, as their natural host and four species have been documented (*Tropilaelaps clareae*, *Tropilaelaps koenigerum*, *Tropilaelaps thaii* and *Tropilaelaps mercedesae*). However, it has now been found that two of these, *Tropilaelaps clareae* and *Tropilaelaps mercedesae*, also infest the Western honey bee, *A. mellifera*. *Varroa* has received enormous attention over the past three decades and, as a result, the importance of other parasitic mite species on bees has been largely ignored. Also, because at present *Tropilaelaps* mites are only found in Asia, their threat has been largely overlooked in Australia. If *Tropilaelaps* were to enter Australia, the United States or Europe, the mite's destructive power would be far greater than that of *Varroa* due to its much shorter reproduction cycle. *Tropilaelaps clareae*, which is frequently found infecting European honey bee colonies in Asia, has

16.11: *Tropilaelaps* mites look very similar to *Varroa* mites.

16.12: *Varroa (L)* and *Tropilaelaps (R)* mites side by side.

16.13: Workers removing excess drones from a colony.

a wide distribution throughout Asia, and from Iran to Papua New Guinea, while other species of *Tropilaelaps* have a much more restricted geographic distribution.

Adult *Tropilaelaps* mites are light reddish-brown and, depending on the species, between 0.7 and 1.0 millimetres long and approximately 0.6 millimetres wide. Like *Varroa*, they need to enter cells containing bee larvae shortly before capping in order to reproduce. Forty-eight hours after cell-capping the mite reproduces, laying three or four eggs, each of which will hatch within twelve hours. Five days after hatching the mite larvae will develop into adult mites.

Tropilaelaps mites depend on softer brood for food since their mouthparts cannot pierce the relatively tough body walls of adult bees. The developing mite feeds on the haemolymph (blood) of capped bee larvae, depriving it of the nutrients required for growth. At the same time as the young adult bee emerges from its cell, the new generation of adult mites (and their mother) leave the cell in search of fresh hosts to live off. Female *Tropilaelaps* mites become pregnant while living in their initial cell and as pregnant females will die within two days of leaving the initial host bee they need to deposit their eggs quickly into a new cell if the eggs are to survive.

SIGNS OF INFESTATION

Tropilaelaps is often mistaken for *Varroa* although distinguishing between the two is relatively straightforward.

- A *Varroa* mite is larger, crab-shaped and wider than it is long.
- *Tropilaelaps* are about a third as wide as a *Varroa* mite, or 1-millimetre long and 0.6 millimetres wide.
- The *Tropilaelaps* mite's body is elongated.
- Adult *Tropilaelaps* mites run rapidly over infested brood combs. Video shot by United States Department of Agriculture (USDA) researchers show the mites running around on the comb at a much faster rate than the *Varroa* mite.
- Unlike *Varroa*, *Tropilaelaps* mites live in brood cells rather than on adult bees.

Infection by *Tropilaelaps* causes abnormal brood development and the death of both brood and adult bees that will cause colony strength to decline, resulting in absconding or the total collapse of the colony. With severe infestations, hives have a noticeable smell of decaying bee remains. Adult bees parasitised by *Tropilaelaps* during development have reduced lifespans, lower body weight, and wing and leg deformities.

European honey bees, *Apis mellifera*, lack the well-developed defences to *Tropilaelaps* mites that the giant Asian bee, *Apis dorsata*, has developed over thousands of years and are therefore more susceptible to the effects of infestation. Unlike *Apis mellifera*, *Apis dorsata* can bite and injure *Tropilaelaps* mites and it is also more efficient at grooming and removing the parasites.

The majority of *Tropilaelaps* mites live and reproduce inside brood cells as they can achieve a longer lifecycle there, living only about three days on an adult honey bee outside of a pupal cell. Perhaps the most alarming characteristic of the mite is its reproduction rate and breeding cycle. One to four adult female mites are generally found in a cell reproducing at one time, however there are reports of mites being found in quantities over three times that many in just one cell. They can quickly outnumber any *Varroa* mites in colonies, hatching in only twelve hours and reaching full maturity in just six days. Within 24 hours of emerging from a brood cell the mites enter another cell and begin reproducing, laying about four eggs per adult female at a time.

SPREAD

The spread of *Tropilaelaps* is similar to that of *Varroa*. *Tropilaelaps* mites move easily and can readily transfer from one bee to another within the colony. To move between colonies they depend upon adult bees for transport through the natural processes of drifting, robbing and swarming. Within an apiary they can be spread through the distribution of infested combs and bees through poor beekeeping management practices. The rapid distribution of the mite to new areas is usually due to the beekeeper transporting colonies in search of new food or for pollination.

The lifecycle of *Tropilaelaps* is similar to that of *Varroa destructor*, but *Tropilaelaps* develops much faster and the intervals between successive reproductive cycles are very short. As a result, if both types of mite are present in the same colony, *Tropilaelaps* numbers can build up at a much faster rate than those of *Varroa*.

TREATMENT

Treatment for *Tropilaelaps* is expected to be similar to that of *Varroa*, although since this mite has only been found in Asia and has not been detected in Australia, the United States or Europe, no approved treatment has been agreed upon. Similar to *Varroa* mite detection in Australia it is expected that government apiary inspectors and beekeeping associations will issue directions for the control and management of *Tropilaelaps* to assist the beekeeper.

An important difference between *Varroa* and *Tropilaelaps* is that a mature Varroa mite can survive for many months outside a brood cell. It does this by attaching itself to an adult bee and digesting its fat body after biting into the host bee's abdomen. *Tropilaelaps* cannot survive this way since their mouth parts are not sufficiently strong to bite into an adult bee's abdomen.

One way in which a *Tropilaelaps* infestation can be controlled is to reduce the hive brood for several weeks. This can be done by either removing the queen from the hive for three weeks or by placing her inside a cage within the hive so that in either instance no more eggs are being laid.

SUMMARY

- The *Varroa* mite is today's most serious problem facing European honey bee beekeepers. Although expensive and disruptive to both honey production and pollination, experience in other countries shows us that mites can be managed, but only if the beekeeper masters significantly different management practices. Again, experience in other countries shows us that once these techniques have been mastered, control of a *Varroa* infestation becomes a routine part of a beekeeper's colony management practices.
- *Tropilaelaps*, a mite currently only found on the Western honey bee in Asia, may cause even greater damage to the beekeeping industry than *Varroa* if it were to escape out of Asia into Australia, North America or Europe. It is expected that if *Tropilaelaps* were to be detected in Australia, management techniques to control the mite would be very similar to those used to control *Varroa*.

17.1

17.2

17.3

17.

Pests of the honey bee

WAX MOTH

Wax Moth can cause severe damage to frames of honey and brood although infestation is usually limited to weaker hives that have insufficient numbers of bees to provide protective cover on frames to fight the moth.

There are two related, although physically different, types of Wax Moth in Australia: the Greater Wax Moth (GWM), *Galleria mellonella*, and the Lesser Wax Moth (LWM), *Achroia grisella*. The Greater Wax Moth is the larger and more destructive of the two with a female adult length of about 20 millimetres and a wing span of between 30 millimetres and 41 millimetres. It is this pest that will be described in this chapter. The Greater Wax Moth is a brown-grey colour and may have a number of speckles on her wings. Infestation with Greater Wax Moth is often called *Galleriasis*. Larvae length when fully grown is up to 28 millimetres.

GREATER WAX MOTH

The female Wax Moth enters colonies during the night when the defences of the hive are at their lowest. Once inside, the moth lays her eggs directly on comb or in other cracks in the hive that cannot easily be cleaned by the bees or seen by the beekeeper. Upon hatching, the larvae initially feed upon base-board debris before burrowing into pollen storage or honey cells. As the Greater Wax Moth larvae develop they burrow towards the centre of the colony where they are protected by the capped cells, leaving behind a trail of destruction in the form of a dense mass of silken webs and faecal debris. Unattended combs can be completely

17.1: The Greater Wax Moth (GWM), *Galleria mellonella*.

17.2: A Wax Moth larva.

17.3: Cocoons in which the Wax Moth pupae live prior to emerging as fully grown moths.

17.4: A tangled mass of silk webbing is a later sign of infestation by the Greater Wax Moth. Black faeces are visible in the silk matting.

destroyed in as short a time as ten to fifteen days, and the larvae will also devour young brood if there is insufficient other food available. Between 50 and 150 eggs are laid in each batch by a single female and the maturation period for Wax Moth larvae, depending on temperature and the availability of food, typically takes eighteen to twenty days in tropical regions and a little longer in temperate zones of Australia. In very cold climates the eggs can remain semi-dormant for up to six months until warmer conditions arrive. The newly emerged larvae are a creamy white colour but as they reach maturity they turn grey. Infestation by the Greater Wax Moth, *Galleria mellonella*, can result in the young brood being unable to leave their cells due to entanglement in the silken threads of the Wax Moth cocoons.

Studies overseas have shown that the newly hatched larvae of the Greater Wax Moth can travel more than 50 metres and are therefore capable of moving to neighbouring hives. The average colony copes well and will control a small number of larvae but if bee numbers are reduced for any reason Wax Moth larvae may take over and destroy the hive.

Careful identification of larvae by the beekeeper

17.5: Damage to a hive caused by Greater Wax Moth larvae.

is necessary because Wax Moth larvae and Small Hive Beetle larvae are similar in appearance. Wax Moth larvae have three pairs of legs on the front of the body and have other uniform pairs of prolegs along the rest of the body. Small Hive Beetle larvae on the other hand only have three pairs of legs at the front of the body and no prolegs along the rest of the body. Wax Moth larvae are soft and fleshy, whereas the Small Hive Beetle larval body is rigid and hard. It is common for both types of pest to be present in the same hive. If larvae are found in the hive a simple way for the beekeeper to determine if Wax Moth is present rather than Small Hive Beetle is the presence of a mess of webbing in and on comb. This is a clear indicator of Wax Moth as Small Hive Beetle do not create either webbing or the characteristic Greater Wax Moth burrowing damage seen in the wooden parts of the hive or frame.

SIGNS OF INFESTATION

For both types of Wax Moth the signs are the same: larvae crawling in the hive. Larvae burrowing through comb, particularly brood comb, are often the first signs of infestation in hives. Cocoons can also sometimes be found on the top of frames or in other parts of the hive. Cocoons are obvious due to their larger size and are often covered in very small black pellets of larval faeces.

Later stages of infestation are very distinctive due to the tangled mass of silk webbing that covers the parts of the frame where larvae are burrowing. When the infestation is even further advanced, the comb in the frame falls apart, a very dense mass of impenetrable silk matting has been spun and larvae can be seen crawling over what is left of the comb.

Greater Wax Moth larvae are responsible for the physical damage to hive boxes and frames that appears as gouges or holes in the wood. Often Wax Moth larvae can be seen crawling out of these holes when disturbed.

The adult Wax Moth is harmless and does not damage hives; it is only the larvae and pupae that cause serious problems for the beekeeper.

SPREAD

Since the main source of food for Wax Moths is found inside a hive or feral nest, the moth will fly from one colony to another in search of a new home or food.

TREATMENT

There are no approved chemical treatments for the Greater Wax Moth within the hive so effective control comes down to the management practices of the beekeeper. Since the moth targets weaker colonies the beekeeper should ensure that the hive remains strong by merging weak colonies. A strong hive is a healthy hive so ensure that there is plenty of pollen or pollen substitute available for the bees and that there is a suitable reserve of capped honey during periods of drought. If the hive has too many boxes so that the distribution of bees throughout the hive is sparse, this will give the moth an opportunity to flourish in the exposed comb. It is therefore advisable to keep empty space within the hive to a minimum by limiting the number of supers added and by monitoring the space required by the bees. A healthy colony will cope well with a small infestation of Wax Moth larvae and the adult bees will quickly remove these. Like many pests and diseases affecting the honey bee, a weak colony will be most vulnerable to Wax Moth larvae build-up and may in time abscond or be completely annihilated.

Wax Moth also cause problems with frames containing used comb, unclean wax or used hive boxes stored in the open or in sheds. To minimise infestation, frames should be stored in airtight plastic containers, preferably in cool rooms. If the beekeeper plans to keep capped or partly capped frames of honey for any period of time any Wax Moth eggs or larvae can be killed by wrapping the frames individually in sealed plastic bags and freezing for two days. After that time the frames may be safely stored in airtight plastic boxes as all the larvae will have been killed.

Wax Moth will only settle and breed in dark locations and knowledge of this behaviour can be used to the beekeeper's advantage. Empty frames should be stored inside supers and stacked after placing sticks between each of the boxes to let in light. A lid is unnecessary, for as much light as possible should be allowed to enter through the top of the stacked boxes.

SMALL HIVE BEETLE (SHB)

The Small Hive Beetle (SHB), *Aethina tumida*, mainly affects weaker colonies as there are insufficient worker bees to remove or drive the beetles from the hive. The adult beetle lives in areas of the hive box inaccessible to bees; for example, under the hive mat or in places less often frequented by bees like the corners of lids and cracks in the wood of the box.

The larvae of Small Hive Beetle are often mistaken for Wax Moth larvae although there are significant differences between the two:

- Small Hive Beetle larvae are smaller than Wax Moth larvae. Small Hive Beetle larvae are around 10.1 millimetres in length and the Lesser Wax Moth larvae are up to 13 millimetres in length.
- Small Hive Beetle larvae have three pairs of legs at the front of their body with no prolegs, or smaller legs, along the rest of the body.
- Small Hive Beetle larvae have a row of spines along their back together with two distinct spines protruding from the rear.

17.6

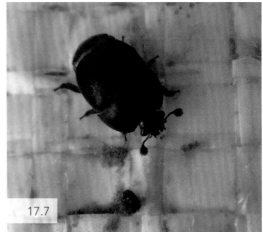

17.7

17.8

17.9

17.6: The Small Hive Beetle.

17.7: A close-up view of the Small Hive Beetle.

17.8: Small Hive Beetle larva. Note the three pairs of legs at the front, the spines on top and the two distinct spines at the back.

17.9: Small Hive Beetle larvae spoiling, or sliming, a frame of honey.

Small Hive Beetle is a relatively new pest to Australian beekeeping, first detected near to Sydney in October 2002. Even at this stage of detection the beetle had become well established around the greater Sydney area and its eradication was deemed impossible. Small Hive Beetle has spread widely since it was first discovered and can now be found in Queensland, New South Wales, Victoria, South Australia and Kununurra (north Western Australia). It was initially thought that the beetle would not move deep into Victoria as the colder climate would not suit its lifecycle, but like many pests it has proved very adaptable and can now be found in even the most southerly parts of the state. Over the next few years we will see beetle numbers rise and spread deeper into Australia. As Small Hive Beetle is still spreading throughout Australia, its full effect on the beekeeping industry has yet to be determined; however, indications of its impact to date in both New South Wales and Queensland show it will increase the workload of hobby and professional beekeepers as well as cause significant economic loss to the industry.

LIFECYCLE OF THE SMALL HIVE BEETLE

Adult female Small Hive Beetles will lay eggs directly onto food sources within the hive, such as pollen or brood combs. Alternatively, female Small Hive Beetles may deposit irregular masses of eggs in tiny crevices or cavities away from the bees. A female Small Hive Beetle may lay 1000 eggs in her lifetime, although some researchers suggest that the number of eggs produced in one female's lifetime might be upwards of 2000. The majority of these eggs hatch within three days.

Newly hatched larvae immediately begin feeding on the available food including honey, pollen and bee brood, though they have a preference for bee brood. Maturation time for larvae is generally between ten and sixteen days. Once the larvae finish feeding, a wandering phase is initiated where male and female larvae leave the hive to find suitable soil in which to pupate. It is believed that the majority of larvae do this at night under cover of darkness.

Larvae in the wandering stage may travel large distances from the hive to find suitable soil. Despite this, most larvae will pupate within 90 centimetres of the hive and will burrow down into the soil less than 10 centimetres in order to pupate. The period of time spent in the ground pupating can vary greatly depending on factors such as soil temperature, moisture and composition. Generally though, the majority of adult beetles emerge from the soil in approximately three to four weeks.

Upon emerging from the ground as adults, the Small Hive Beetle flies in search of honey bee colonies and will probably identify potential host colonies by a range of smells emanating from the hive including honey and pollen. The beetle flies before or just after dark and, upon locating and entering the host colony, will seek out cracks and crevices where it can hide from the bees' often aggressive attempts to evict it.

The mating behaviour of the Small Hive Beetle is not fully understood although it is when the female adult enters the hive and lays her eggs that the lifecycle begins again.

The turnover rate from egg to adult emerging from the soil can be as little as four to six weeks; consequently, there may be as many as six generations in a twelve-month period under moderate climatic conditions.

Small Hive Beetles are a problem in hot, damp climates like Queensland and coastal New South Wales, but are also spreading into the colder regions of Victoria and South Australia.

Even though there may not be any Small Hive Beetles in the drier areas of Australia, if discarded comb, slum-gum or residue from cleaning hives is left on the ground it may become sufficiently damp to host the beetle. Already the Small Hive Beetle is proving more adaptable to different climatic conditions than was initially thought and there is no reason to believe it will not continue to adapt and spread to other areas of the country.

SIGNS OF INFESTATION

Depending on how frequently hives are inspected, the following will give either a clear or fair indication that Small Hive Beetle is present.

- Upon opening an infested hive, beetles are seen scurrying away out of sight, perhaps from on top of the hive mat.
- For the less frequently inspected hive, noticing a slime oozing out of the hive entrance and a generally putrid smell similar to that of rotting oranges. The Small Hive Beetle itself is not the cause of this destruction but its larvae. As the larvae grow they burrow their way through the comb. They prefer comb that has contained brood or pollen. During the burrowing through honeycomb, a residue of yeast is left in the larvae excreta that quickly causes the honey to ferment, then to slime and fall to the base of the hive as a revolting slurry.

Either of these situations is a sign of an established infestation of Small Hive Beetle that will need management. Other observable indications and checks for suspected Small Hive Beetle infestation include:

- Small larvae burrowing through capped comb, both honeycomb and brood comb. Since the comb is capped the larvae may be difficult for the beekeeper to notice during a routine hive inspection.
- Beetles inside non-capped areas of the comb. If unsure, tap the frame onto the lid and observe if any beetles fall out.
- Inspect under the lid of the hive, on top and underneath the frames to see if there are any beetles there.
- Pick up the bottom brood box and check if there are any larvae or beetles on the base board, particularly if the base board is dirty as the beetles may be hiding in the debris.
- If the brood box is attached to the base board, remove all the frames from the brood box and very carefully inspect the base board from the top.

- Check all the nooks and crannies in the hive where beetles may be hiding.
- If plastic frames are used check that beetles are not hiding in the hollow ribs of the frames' sides.
- Inspect metal queen excluders, particularly those with a solid strip edge, as the folded metal can provide a suitable hiding place.
- Remove a super with frames and place it upside down on a lid. Wait for several minutes before returning the super to the hive and quickly inspect the lid for any beetles that may have congregated to hide from the light.

Once the beekeeper easily observes several beetles in a hive that is usually a good indication that there are many others and measures need to be taken to help the bees to keep the pest at controllable levels.

SPREAD

The beetle is attracted to the smell of honey and flies from one hive to another. The migratory nature of commercial Australian beekeeping has ensured that the beetle can easily be shipped around the country, infesting new hives and areas wherever it is introduced. It is believed that the beetle can fly up to 10 kilometres and thus a single infested hive can infect other colonies for an area of over 300 square kilometres. Since feral hives also harbour the beetle, movement of commercial hives into new areas can inadvertently lead to infestation.

TREATMENT

As previously noted, strong hives are reported to be less susceptible to Small Hive Beetle than weaker hives although this has been disputed by some experts, particularly in areas where the beetle has gained a good foothold. In order to obtain stronger colonies, merge weaker colonies or add a frame or two of capped brood to the weaker hive. With either of these two options make sure that you are not transferring any disease from an infected colony.

Small Hive Beetles are attracted to the smell of honey and use it to guide them to hives. As a consequence, to reduce the chance of the beetle being attracted to your hive by the smell, minimise the number of times you open the hive and keep the area around your hive clear of old honeycomb and debris that may act as an attractant. Small Hive Beetles will also fly with swarms so by catching a swarm you may unwittingly introduce the beetle into your apiary.

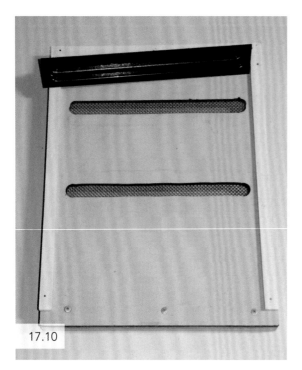

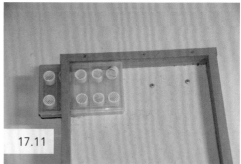

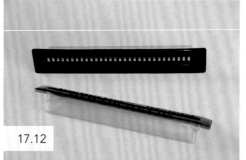

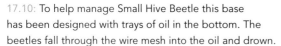

17.10: To help manage Small Hive Beetle this base has been designed with trays of oil in the bottom. The beetles fall through the wire mesh into the oil and drown.

17.11: Another design of base for trapping Small Hive Beetle. The removable box at the back of the base contains oil that the beetles fall into and drown.

17.12: A trap filled with oil that is inserted between frames in both the brood box and the supers.

17.13: Another method of managing Small Hive Beetles is to use a corrugated case filled with the miticide Apithor. This trap is called the Apithor Hive Beetle Harbourage.

17.14: Filling a Small Hive Beetle trap with oil.

The beetles prefer to breed in a humid environment, so in order to keep the inside of hives a little drier, open the lid by a very small amount for a few hours to allow any moisture to escape, say by placing a twig under one side. Leaving the hive like this for longer periods will encourage robber bees, wasps or even more beetles to enter and this option can be time consuming so is really for the hobby beekeeper with time and a small number of hives.

Hive boxes should be kept in good condition; any cracks need to be repaired so that they cannot be used by the beetle to enter the hive and to hide within. Also ensure good hygienic practices in and around the hive by keeping the area clean and tidy and clearing any burr comb from inside the hive lid and removing waste material from the hive floor. This latter practice is made a lot easier if base boards are not fixed to brood or bottom boxes.

Minimising the movement of the hive is also a good idea since by changing its location the beekeeper may be moving it close to a colony that already has the beetle and will unwittingly aid it in its spread.

To minimise the contamination of stored frames by the beetle, if possible keep them in a cool room below 10° Celsius.

Beekeepers should contact their state's apiary inspector to determine which chemical treatments are approved for use in the hive to manage infestation.

Apart from the above-mentioned general management practices, the following techniques have been successfully used to help colonies control the beetle.

- Beetle traps can be placed inside the hive between frames. There are several types of beetle trap for sale through suppliers in Australia. A black-coloured plastic reusable model with a clip-off top is available as is a plastic disposable model with a black trapping surface and a deep clear reservoir. There is also a similar metal design available. The traps work on the principle that the bees chase the beetles into the slotted top of the trap where they drop into the reservoir and are killed either by drowning in a mixture of vegetable oil and a touch of apple cider vinegar or are smothered by garden lime or diatomaceous earth. Any kitchen vegetable oil can be placed in either trap and this option has the advantage that the oil mixture is non-toxic, does not have an unpleasant taste and will not contaminate honey if accidently spilled onto the frame. The disposable traps tend to have a deeper reservoir than the reusable models and have the additional advantage that the clear plastic reservoir makes it easy to do a dead beetle count and to check the level of oil.

- Another popular method used to trap beetles is to use carpet underlay or felt-backed linoleum as a hive mat. The mat fibres trap the beetles and these can be disposed of during the next inspection of the hive by the beekeeper. A disadvantage of this method is that often the odd bee or two will also be caught in the felt.

- Disposable kitchen wipes such as Chux Wipes and some types of carpet have also been shown to be effective in trapping Small Hive Beetles — success can often depend on the beetle numbers within a particular hive and how vigorously the bees pursue them.

- Corflute, a type of plastic with a hollow, tubular core, can be inserted into the hive entrance or placed under the lid. The beetles will use the hollow tubes of the corflute to hide from the bees and the corflute can be removed by the beekeeper and destroyed. To ensure that the corflute can be conveniently inserted and removed from the hive entrance a piece of stiff wire is attached to it so that it can be pushed in or pulled out easily. An effective variation of corflute is a design by the New South Wales Department of Primary Industries and the Rural Industries Research and Development Corporation (RIRDC) that impregnates a hollow-fluted matrix with a miticide. The device is called the Apithor Hive Beetle Harbourage. The Apithor trap is pushed through the hive entrance and onto the base board of the hive. The beetles enter the hollow matrix to escape the bees and are killed by the miticide in the trap. This device has been shown to be effective in controlling beetles and the trap is small enough to be easily inserted and removed through the hive entrance. The miticide is only on the inside of the trap matrix where the mites hide so cannot be touched by the bees.

- Another method of control is to use a specially constructed base that has a tray of vegetable oil, garden lime or diatomaceous earth with wire gauze above the tray so that the beetles can climb through the wire but the perforations are too small for the bees to enter. The beetles fall into the tray where they are killed and can be cleaned out by the beekeeper. For ease of use, the tray needs to be removable from the base without disturbing the rest of the hive. If the hive needs to be dismantled to remove the tray of dead beetles this will mitigate many of the advantages of the technique.

The design of traps for the Small Hive Beetle is limited only by the imagination of the beekeeper. Those with an inventive mind have successfully used empty CD cases, fishing tackle boxes and empty plastic screw containers as inexpensive, but effective, beetle traps. My own preference is for traps that contain an oil and vinegar mixture, as these seem to offer no adverse effects to bees or honey and are easy and inexpensive to install. I prefer the disposable traps with a deep clear reservoir as they are easy to refill, and I am able to monitor dead beetles with a quick visual check. The deep reservoir of the trap also makes it less likely that the contents will spill if the hive is moved. Although sold as disposable the traps can be used for long periods. Some bees have a tendency to propolise or wax up the top entry point of traps and you may need to keep an eye on this.

BRAULA FLY

The scientific name for the Braula fly is *Braula coeca* and is often incorrectly termed the Bee Louse. The fly is found in Tasmania but not in other parts of Australia.

The Braula fly is unusual in that it is wingless, flattened and lives in honey bee colonies by holding on tightly to the body hairs of adult bees, usually on the head, using a set of comb-like structures on its front legs. When hungry, the fly moves to the mouth of the bee and steals some of the food being fed to it by other bees. Adult Braula flies are reddish-brown in colour and are regularly mistaken for *Varroa* mites.

The Braula fly does not damage or parasitise any stage of the honey bee lifecycle and is not usually considered a significant pest or threat to colonies. It is worth noting, however, that if Braula is living on a queen it may reduce the amount of available food, thereby resulting in an impaired ability to lay eggs.

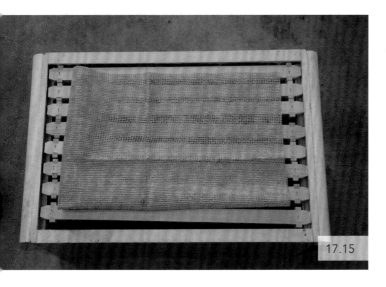

17.15: An inexpensive way to trap Small Hive Beetles is to lay a Chux cloth on top of the frames under the lid instead of a hive mat.

17.16: A Braula fly.

17.17: Braula flies on a bee's thorax.

The main economic impact of Braula fly occurs as a result of the fly laying eggs on capped honeycomb and the larval stage burrowing into the cell under the honey cappings. The sight of Braula larvae burrows in comb honey in particular can detract from its visual appeal for sale. Liquid honey is not affected by Braula, since most honey is extracted mechanically and will not be affected by the fly since it will be filtered out during the extraction process.

SIGNS OF INFESTATION

Braula flies are sufficiently large that they can be seen with the unaided eye on adult bees including the queen. Simple observation of adult bees, particularly the resident queen bee, may reveal the presence of adult Braula flies. Interestingly, tobacco smoke causes the Braula fly to fall off the bee so the useful addition of a few grams of tobacco to the smoker will cause the fly to fall to the floor of the hive. A sticky sheet of paper on the hive floor will catch any fallen Braula where they can easily be seen by the beekeeper.

TREATMENT

The use of small amounts of tobacco in the bee smoker has been demonstrated as a very effective means of killing adult Braula flies. However, continued exposure to tobacco smoke will affect bees and may lead to some adult bees dying.

Comb honey should be stored in a freezer for two days after its removal from a beehive. This will ensure that all stages of the Braula fly larval lifecycle are killed. This will also serve to kill all larval stages of other beehive pests such as Wax Moth and Small Hive Beetle.

SPREAD

Adult Braula attach themselves to adult bees and can be spread by swarms, drifting bees, and queen bees. Overseas, Braula fly infestation is less common, possibly due to the chemical treatment of colonies for *Varroa* which would also limit the incidence of Braula infestation.

SUMMARY

There are several pests of the honey bee that, like Small Hive Beetle, can be very destructive of the colony, or, like Braula fly, are only minor irritants. The main pests are:

- Wax Moth — which comes in two species, the Greater Wax Moth and the Lesser Wax Moth. The Greater Wax Moth is the more destructive of the two.
- Small Hive Beetle — recently introduced into Australia, this beetle is causing serious economic harm to the beekeeping industry and is spreading further north and west and into the colder regions of Victoria and Tasmania.
- Braula fly — found only in Tasmania, this is only a minor irritant to the honey bee and is not a cause for concern.

Currently the limited use of a Fipronil-based insecticide has been approved against Small Hive Beetle in the form of the Apithor Hive Beetle Harbourage.

18.1 18.2

18.

Other problems of the honey bee

PESTICIDE AND FUNGICIDE POISONING

A common cause of death or poor health in a colony is the ingestion of pesticides or fungicides by bees. Spraying flowers or orchards with pesticides or fungicides, both commercially and domestically, can lead to the death of a colony that collects nectar or pollen from that area. Typically a large ingestion of poison by the bees in a colony would be shown by many bees lying dead or dying outside of the hive. Although this is a worst-case scenario, it is a relatively common occurrence, particularly with large-scale crop spraying. Landowners or contractors who do not inform local beekeepers about planned spraying operations may be held responsible for the economic loss of the beekeeper's colonies.

A more common and perhaps more insidious scenario is the low-level absorption of pesticides or fungicides by bees at levels below those that are immediately obvious and evidenced by large-scale die off of bees. Sub-lethal poisoning — that is, a dose many orders of magnitude below a lethal dose — can result in poor brood laying by the queen, as well as insufficient collection of nectar or pollen by foragers. Also, as even low-level chemical poisoning weakens the immune system of bees this can result in a compromised immune system and susceptibility to higher levels of disease, for example, *Nosema*. A further demonstrated effect of sub-lethal doses of some pesticides is found in returning foragers unable to find their way back to the colony, with up to one-third failing to return to the hive for this reason.

18.1: Bees killed by pesticides outside a hive.

18.2: Bees showing the characteristic extended tongue or proboscis common with pesticide poisoning.

Chemical pesticides may be applied topically to plants as a surface spray and in this situation bees absorb the chemical when they either crawl over the surface or collect nectar or pollen from affected plants. Other types of chemical insect control readily available are the systemic pesticides which, when applied, are absorbed by the plant into its cellular system and tissues. This can occur through the ground, by topical spraying or by direct incorporation into seeds. The plant, its pollen nectar and any excreted water droplets or resins carry the pesticide and unsuspecting bees forage and return to the hive with the chemical.

In the last ten or so years another class of insecticide chemicals has been developed and advanced: the neonicotinoids. These are systemic pesticides and were further developed as they were shown to have low levels of toxicity in mammals. Neonicotinoids are a class of neuro-active insecticides chemically similar and modelled on the natural insecticide nicotine. They work by blocking a specific neural pathway that is more abundant in insects than in mammals, resulting in selective or targeted toxicity of insects. Neonicotinoids work on the insect's central nervous system causing overstimulation, eventual paralysis and then death.

One of the available branded neonicotinoids, Imidacloprid, has become the most widely used insecticide in the world and there is a concern that neonicotinoids may have some connection with the health problems honey bees are experiencing in the United States and parts of Europe. There is certainly fear among beekeepers that, at the very least, the use of this class of chemical may result in a culmination of many and varied problems for the honey bee.

In January 2013, the European Food Safety Authority stated that neonicotinoids pose an unacceptably high risk to bees and questioned the previous science relied upon by regulatory agencies to claim the safety of these pesticides. In response to this and the report of the Food Safety Authority in May 2013 the European Commission adopted a proposal to restrict the use of three pesticides belonging to the neonicotinoid family for a period of two years. The proposal restricts the use of these for seed treatment, soil application and also the foliar treatment of bee-attractant plants and cereal crops.

The Commission has stated that as new information becomes available it will review the conditions of approval of the three neonicotinoids and take into account any relevant scientific and technical developments. It is interesting to note that exceptions to the possibility of treating bee-attractant crops in closed greenhouses and open-air fields after flowering were contained in the proposal.

As with any controversy, it may be many years before there is sufficient data on the real effects of this class of pesticide on honey bee health and populations. Any low-level infection or agent can, when added to other environment stresses, be the tipping point to the compromised health and loss of viability of a colony and the fact that these insecticides were developed to target insects, of which the honey bee is one, will continue to cause beekeepers warranted concern.

In the United States large numbers of hives are routinely shipped around the country for pollination purposes, the bees spending long periods foraging in various monocultural crop situations with feeding supplemented by high fructose corn syrup. Given the ever-increasing number of diseases treated by numerous chemicals, plus the various mites including Tracheal Mite and *Varroa destructor*, and pests like the Small Hive Beetle, plus the accumulating chemical build-up within hives, it is not too big a jump to understand concerns that the added burden of a neonicotinoid has pushed the bee to the point where its immune and other systems are severely compromised.

For the beekeeper, the best safeguard against accidental exposure of bees to lethal pesticides and fungicides is to develop an awareness of whether and where these chemicals are being used near hives. A diplomatic approach to a nearby user, who may be unaware of the affect the chemicals sprayed are having on nearby hives, may result if nothing else in their use during more bee-friendly times of the day or at least give the beekeeper an opportunity to either close or move hives during the period of spray drift.

CHILLED BROOD

A healthy colony containing a large number of workers is able to keep its brood at 34° Celsius to 35° Celsius. This is the optimum temperature for raising brood and this temperature needs to be well regulated if the colony is to remain healthy and to produce viable brood. If, for some reason, temperature regulation fails, the developing brood may become chilled to the point where development stops and death occurs.

There are several causes of chilled brood including:
- smaller hives or nucleus that have insufficient numbers of workers to keep the colony warm
- hives that are rapidly expanding during spring that contain large

numbers of brood but do not yet have sufficient numbers of adult workers to keep the brood warm

- hives that have lost large numbers of adult bees due to disease or pesticide poisoning
- beekeepers opening hives in cold weather and allowing warm air to escape.

The best way to minimise chilled brood is to keep strong colonies and not to open hives in cold weather. Another technique is to reduce the size of the entrance during winter months to minimise the amount of cold air flowing through the colony. Fortunately in Australia there are few areas where it is sufficiently cold for long enough to cause real concerns about chilled brood.

OVERHEATING OF THE HIVE

A more common problem in Australia are hives overheating due to hot weather or to fire. Daytime temperatures in many inland parts of the country reach over 40 °Celsius during summer and this can prove detrimental, stressful and at times fatal, particularly if there is insufficient water nearby for the bees to cool the colony. To minimise the effect of heat, the following practices have proved to be effective.

- Where possible, locate hives so that they are in shade during the hottest part of the day usually between 10 a.m. and 6 p.m.
- Keep the entrance to the hive as open as possible.
- Do not open the hive when the air temperature is above 33° Celsius.
- Ensure air vents at the hive lid are adequate and are unblocked to enable the fanning bees to keep cool air flowing through the hive.
- Try using seven frames in an eight-frame super or brood box or nine frames in a ten-frame box, to allow for greater air circulation around the frames.
- If possible, keep hives near larger bodies of water as this will both cool the air as well as provide drinking water for the colony. For the hobbyist with a few hives there are additional possibilities. The top of hives can be watered before it becomes too hot, taking care not to drown any bees. This and watering under and around the boxes frequently during extreme weather conditions will assist the bees' own efforts inside the hive where collected water is fanned with numerous wings in order to produce a cooling air-conditioned effect. Other hobbyist beekeepers place old beach umbrellas over their hives and water the top of these in order to offer some relief to bees in the hives below.

- When transporting hives, make sure that there is a good airflow through the hive and that the entrance is not blocked off. Perforated hive-entrance closers can be used to prevent bees escaping while allowing air to pass in and out of the hive. The use of a screened travelling top is recommended as, again, it allows air to circulate and provides the bees with fresh air.
- Paint hives with white or other light-coloured paint as darker colours attract and hold heat.
- Provide insulating shade covers on top of the normal hive lid, particularly if this is metal.

Since fire is a major concern to beekeepers in many areas, where possible keep hives away from long grass or vegetation that can burn. Reduce vegetation to a minimum around hives particularly during hot and dry periods.

Bees are affected by the heat and they can become very aggressive and difficult for the beekeeper to manage without good protective clothing and copious amounts of smoke. Obviously it is not a good idea to work hives in hot weather as this destroys the interior temperature the bees have fought so hard to maintain and will take a considerable time for them to re-regulate. If the beekeeper is aware a hot day is forecast and it is necessary to open the hive this should be done in the early morning when it is cooler or, if absolutely necessary, in the late evening when of course most of the bees will be inside and ready to object strongly to the interference. The best option is to leave hives alone until the temperature drops, hopefully after a few days. The beekeeper should always be aware that a total fire ban may be in place on very hot days and inspections will need to be done without the benefit of a smoker. Some beekeeping supply stores sell a liquid smoke that can be mixed with water and sprayed over the bees using a spray bottle. Although not as effective as smoke from a smoker, a spray or two of liquid smoke will help to calm the bees if the hive needs to be inspected on total fire ban days.

DAMP HIVES

The classic sign of a damp hive is condensation on the inside of the lid. The condensation will pool into droplets which will drop down onto the bees in the hive. Dampness in hives contributes to or causes a lot of health problems in bees, including *Nosema* and Sacbrood.

If dampness in hives is a problem try doing the following:

- Add ventilation at the top of the hive either by providing air vents in the lid or by drilling 2.5-centimetre holes at the top of the uppermost super and covering with vents. Another technique is to place lollipop sticks, twigs or small stones under the lid rim, raising the lid by about 3 millimetres, thus providing greater airflow throughout the hive. A gap any larger may open the hive to predators or robbers. See also Chapter Seven on the procedure for the first inspection of spring.

- Make sure that the base of the hive slopes forward so that any condensation flows out through the entrance and does not pool at the back of the base.
- Keep the hive away from damp ground and preferably raise it off the ground about 30 centimetres.
- Place the hive in a sunny but not a hot location.

Dampness mainly occurs during the winter when the bees respirate in order to keep the colony warm. Water is exhaled during respiration that then collects inside the hive in the form of damp lids, walls and bases. Adding additional ventilation during cold periods to allow the moisture to escape will be unhelpful, as this will cause cool air to enter the hive and the bees to respirate more intensely in order to keep the colony warm.

18.3: A cane toad waiting at a hive entrance for exiting and returning bees.

18.4: Raising hives above the ground about 60 centimetres will deter cane toads from attacking the foragers.

CANE TOADS

The cane toad, *Rhinella marina* (formerly *Bufo marinus)*, an introduced species, can now be found widely over northern Australia, including New South Wales, Queensland and the eastern parts of the Northern Territory. Cane toads generally feed at night, spending the day hiding under logs, under hives or in small holes. Cane toads have become a major pest to the beekeeper as they sit outside a hive at night and devour many workers from around the hive entrance.

Adult cane toads are usually heavily built and weigh an average of up to 1.8 kilograms. Their size may vary between 15 to 23 centimetres and their skin is warty. The colour on their back and sides varies from olive-brown or reddish-brown, grey, and yellow while their bellies are semi-yellow or semi-white with darker mottling. Their body is round and flat with prominent corneal crests and light mid-dorsal stripes.

18.3

18.4

While their front feet are unwebbed, their back feet have tough, leathery webbing.

The easiest way for the beekeeper to minimise losses from cane toads is to raise the base of the hive about 60 centimetres above the ground. A variety of stands can be used for this purpose. Fixed stands are good if there is no plan to move the hive, but for the migratory beekeeper stands that can be easily assembled or taken down are best.

ANTS

There are many types of ant in Australia and they can be a significant pest by robbing honey, pollen, brood and even adult bees from the hive. Colonies that have been or are being robbed by ants may become aggressive and those that have been seriously and

18.5

18.5: Coastal brown ants.

18.6: Meat ants.

18.7: Sugar ants.

18.8: Green tree ants.

18.6

18.7

18.8

repeatedly attacked will become depleted and stressed with a resultant reduction in honey production.

Ants often become a problem for the beekeeper when sugar syrup is being fed to colonies and in this situation may invade the hive or syrup feeder in large numbers.

There are two small ants that cause problems to beekeepers in the northern parts of Australia: the small black ant and the coastal brown ant.

There are several larger ants that may also be a problem for the beekeeper and these include meat ants, sugar ants and green tree ants.

Ants' nests can be destroyed by the application of chemical insecticides either in powder form or as a solution. Green ants, which make their nests high in trees, can be destroyed by applying insecticide around the base of the tree or by injecting insecticide directly into the nest. There are many insecticides approved for this purpose and the beekeeper should contact a local agricultural or beekeeping supplies store. Alternatively the nearest state apiary inspector will be able to provide the name of insecticides that have been approved for use in the local region.

A common physical barrier for ants is to modify the cane toad stand mentioned on p. 103 and 293. The legs of the stand can be made impassable for ants by tying a cloth around them covered in sticky oil or grease. Insecticide applied carefully so as not to kill the bees may also be used but the insecticide needs to be reapplied more frequently than oil and may pose a risk to the hive of bees housed on the stand. Ants will be able to enter the hive even if it is fully protected by a physical barrier if long grass or other vegetation is allowed to touch the hive or parts of the stand above the physical barrier as they will use these as a bridge to gain access to the hive.

MICE

Generally mice do not appear to be a large problem for the beekeeper in most parts of Australia. Bee equipment suppliers often sell mouse guards which attach to the front entrance of the hive, and from my many years of experience in my own beekeeping business, I find that customers in Victoria generally purchase entrance guards as a precautionary measure rather than to deal with specific mice problems. There is no doubt that the average beehive is an attractive proposition for a mouse looking for warm winter quarters and it is true that in some parts of the country mice may be a problem towards the end of autumn or during the winter. A mouse will eat comb and generally make a mess in a hive, dragging in straw and other nest material. Often the colony seems to ignore a live mouse in the hive even though the bees are more than capable of killing it. In instances where the colony does attack and sting to death an intruding mouse they are unable to remove it from the hive due to its large size. Not to be outdone the bees cover the dead mouse with antiseptic propolis to entomb the

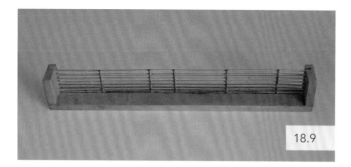

18.9

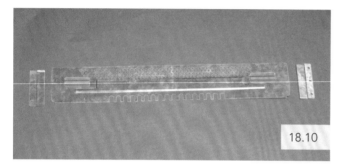

18.10

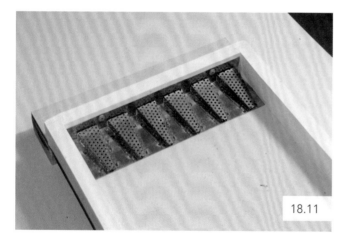

18.11

18.9: Mouse guard to place over the hive entrance.

18.10: A combined mouse guard and entrance closer with attaching brackets.

18.11: Base with an internal mouse guard protecting the entrance.

carcass and thereby prevent the spread of disease into the colony from the decaying body.

As previously mentioned, specialised mice guards are available from some bee equipment suppliers to cover the entrance to hives. These are usually made of a galvanised metal and have a number of small holes to allow entry and exit from the hive by the bees but are too narrow to allow a mouse entry. If the beekeeper has or foresees that there may be a problem with mice entering hives and wishes to do something to narrow the hive entrance it is worth noting that a mouse is able to squeeze through a hole about the size of a pen, that is, approximately 6 millimetres in diameter to get at food or to make a home. Mice are able to do this because the bones in their head are relatively soft and additionally they have a slender, flexible body that can easily fit through a very small entrance. If mice are a problem, consider putting a mouse trap either outside or under the hive, suitably protected from the elements and from domestic and other wildlife. This may go part of the way to fixing any problems that may be caused to the colony by this inquisitive rodent.

COLONY COLLAPSE DISORDER (CCD)

Before discussing Colony Collapse Disorder it is important to note that a collapsed colony and Colony Collapse Disorder (CCD) are not the same thing. Colony Collapse Disorder is a very general disorder found in the United States, the causes of which are poorly understood, and it was regarded as being a major problem to American beekeepers. A collapsed colony, however, is more specific and can be caused by a variety of often-identifiable ailments affecting the bees in a particular colony.

Once Colony Collapse Disorder was first identified in the United States in November 2006, there was a concerted effort to identify a single cause for this large yearly loss of colonies, sometimes reaching as high as 30 per cent of all American hives lost during the winter months. The US Department of Agriculture (USDA), however, now acknowledges that CCD is not caused by a single pathogen or cause but by a range of issues some of which are man-made. These include:

- Infestation by *Varroa*, the most harmful affliction of the honey bee, is closely associated with high winter losses of colonies.
- Multiple virus species have now been found in both bees and *Varroa* mites, including Deformed Wing Virus (DWV), Israeli Acute Paralysis Virus (IAPV) and Black Queen Cell Virus (BQCV), and are believed to be significant factors in poor colony health.
- *Varroa* is known to increase the levels of viruses in bees, as it acts like a mosquito injecting viruses into the brood during feeding.
- The bacterial disease European Foul Brood (EFB) is being detected more frequently in colonies and may be linked to colony loss.
- Nutrition has a major impact on colony health and longevity. The increased use of monocultures, where a single crop is grown over very large areas and bees transported to the crop for the purpose of pollination — for example, almond pollination — is proving harmful to bees due to the lack of a balanced diet and the absence of any supplementary forage within the bees' flight range.
- Gut microbes used by the honey bee to break down pollen, fight some gut diseases, digest food, transport nutrients from the gut into the body and for detoxifying chemicals are being harmed by the increasing use of pesticides, miticides and fungicides which the bees unintentionally digest.
- Transporting hives sometimes many thousands of kilometres each season for pollination has been shown to be detrimental to bees' health.
- High fructose corn syrup (HFCS), which is used widely in the United States as a food supplement, has been shown to be not as effective as honey in developing and keeping healthy the microbes within the gut of the bee.

- The use by local councils and domestic home owners in the United States of herbicides to eliminate 'unsightly' broad leafed weeds on grass lands, parks and lawns is depriving bees of both valuable food and further adding toxins to their bodies.
- Chemicals used within the hive to manage *Varroa*, Small Hive Beetles and Tracheal Mite are proving more of a problem to bees than a solution to pest control.
- Acute and sub-lethal effects of pesticides on bees have been increasingly documented and are a major concern to beekeepers. In some situations, two apparently harmless chemicals that are applied together prove to be significantly more harmful to bees than either chemical applied singly. These effects, together with the long-term cumulative build-up of apparently harmless chemicals, are not well understood and need to be further researched before we are able to discuss with confidence any harm they are or may cause to bees and other living things.
- Poor management practices by the beekeeper.

Although the above list represents what scientists today believe are the major contributing factors for a colony's collapse to be classified as being due to CCD, not all of the above need to be present for there to be irreparable harm to an individual colony and for the collapse to be classified as being due to CCD. Many of the contributing factors listed above come from issues that are unrelated to each other. As a result there will not be a single silver bullet that will eliminate CCD. The solutions to the problem will need to come from diverse areas and will involve many groups of people from industry, beekeepers, government and academic areas working together, and who could even today spend more time talking with each other in an effort to minimise the causes of large-scale bee deaths.

SUMMARY

There are a number of general issues that may adversely affect your colony, although most can usually be easily dealt with by the beekeeper. Pesticide and fungicide poisoning is probably the issue that the beekeeper has the greatest difficulty managing. This is often because the beekeeper is unaware that chemicals are being used in the vicinity of hives and the people who apply them may not be aware of or give proper consideration to honey bee colonies in their area. Apart from pests and disease the more common general problems experienced by Australian beekeepers are:

- pesticide and fungicide poisoning
- chilled brood
- overheating of the colony
- damp hives
- cane toads
- ants
- mice.

19.1

19.2

19.3

Other types of bees and wasps

So far I have discussed the European honey bee, *Apis mellifera*, as well as some of the most important species of native bees found in Australia. There are also established populations of the invasive Asian honey bee (*Apis cerana*), the Buff-tailed bumblebees (*Bombus terrestris*), the solitary African Carder bee (*Afranthidium (Immanthidium) repetitum*) and two species of wasp. Wasps are often mistaken for honey bees, but differ fundamentally in their appearance, behaviour and nesting biology. Also, globally there are other species of honey bee, as well as hornets that, if they were to enter Australia, could pose a significant biosecurity risk.

THE ASIAN HONEY BEE (*APIS CERANA*)

Globally there are twelve currently recognised species of honey bee. One of these, the Asian honey bee, *Apis cerana*, is the most widely distributed honey bee in Asia, found natively from northern Russia and Japan, to Pakistan and Indonesia and everywhere in between. It is extensively used in Asian countries to produce honey, wax and also brood comb, which is eaten as a source of food. *Apis cerana* has many common names in English including the Asian honey bee, Asiatic bee, Asian hive bee, Indian honey bee, Indian bee, Chinese bee, Eastern honey bee, Mee bee and fly bee.

Across Asia, *Apis cerana* can be found at higher altitudes, at sea level, in tropical, sub-tropical and temperate regions, in deserts and in areas where there is high seasonal rainfall. The Asian honey bee has been extensively spread by human activity, most recently by

19.1: The Asian honey bee (*Apis cerana*) workers on brood comb.

19.2: *Apis cerana* worker.

19.3: *Apis cerana* bees on a frame of brood comb in Vietnam.

19.4: Hives of *Apis cerana* in Vietnam where the bee is commonly kept domestically for honey production.

missionaries to Papua New Guinea and the Solomon Islands in the 1970s, and the Cairns region of Australia in 2007. As a result of this widespread distribution, different races of the bee have very different temperaments and appearance, leading to the eight currently recognised subspecies of *Apis cerana*.

Although *Apis cerana* has its home across Asia, the particular race of *cerana* found in Cairns in Queensland is *Apis cerana javana*. This race of *cerana* is also found in the areas of Java, New Guinea and the Solomon Islands. *Apis cerana* moved into the Solomon Islands in 2003 and quickly destroyed the limited local

19.5: Hives of *Apis mellifera* also in Vietnam. Notice the white sacks kept on top of the hive that help keep the colony cool.

European honey bee industry. In May 2007, *Apis cerana* was discovered in Cairns. The initial colony may have entered Australia hidden in the mast of a boat that had recently arrived from Papua New Guinea. By the time apiary inspectors became aware of the incursion the colony had swarmed and multiple colonies of the bee were located living in and around Cairns. Initial concern was due to the possibility that the bee would destroy the Australian beekeeping industry and a concerted effort to eradicate the bee was made by both Queensland's Department of Agriculture, Fisheries and Food (DAFF) and by Australian beekeepers. Despite these efforts *Apis cerana* was declared endemic to Australia in 2011 when it was decided to be too difficult to eradicate. No more resources are currently being allocated to the eradication of *Apis cerana* or efforts made to contain it within the Cairns region. At this point in time initial concerns about *Apis cerana* have not been realised as the Asian honey bee found in Cairns does not appear to be significantly affecting local European honey bee colonies apart from the fact that they are both competing for the same floral resources.

To date, most honey bee research globally has been conducted on the European honey bee subspecies of *Apis mellifera*, and much less research has been conducted on the other eleven species of honey bee. *Apis cerana* is the best studied of the Asian honey bee species but there still remains a lot to be learnt about this species. This is particularly true for the race of *Apis cerana javana* found in Cairns. The probable reason *Apis cerana javana* has not been well studied is that it does not produce and store much surplus honey and is of very little economic importance in the areas where it can be found. As a result, the long-term impact of this race of bee in Australia is not fully understood. The effect of its increasing numbers on the European honey bee is only a best guess from limited knowledge of other races of *cerana* in other parts of Asia, although a recent study conducted by the University of Sydney may give some indication of the future.

The study showed that in the Cairns area, about 33 per cent of European honey bee queens had mated with at least one *Apis cerana* drone. A part of the study also compared European and *Apis cerana* bees in China and found that 14 per cent of European honey bee queens had mated with at least one *Apis cerana* drone. Conversely there were no cases found of an *Apis cerana* queen mating with a European honey bee drone, possibly due to the large size of a European honey bee drone and the smaller size of an *Apis cerana* queen. Although this study showed that *Apis cerana* drones can and do mate with European honey bee queens, the resulting eggs were infertile and did not survive to adulthood. The effects of interspecies mating will therefore reduce the number of fertile eggs that a European honey bee queen will lay and must, as a result, produce weaker honey bee colonies.

Morphologically, both *A. cerana* and the European honey bee are very similar in appearance and only an experienced observer is able to tell the difference between bees of the two species. The main differences are outlined in the following table.

CHARACTERISTIC	THE ASIAN HONEY BEE *Apis cerana*	THE EUROPEAN HONEY BEE *Apis mellifera*
Size of body	About two-thirds the size of *mellifera*	
Foraging range	Typically less than 500m	up to 5km

CHARACTERISTIC	THE ASIAN HONEY BEE *Apis cerana*	THE EUROPEAN HONEY BEE *Apis mellifera*
Size of colony	Approx. 2600 in Cairns	50,000 to 60,000
Potential as a crop pollinator	Unknown	Excellent
Aggressiveness	Less aggressive than *mellifera*	
Resistance to disease	Unknown	see Chapter 15
Resistance to *Varroa*, *Tropilaelaps* and Tracheal Mite	*Cerana* and these mites can co-exist without any significant impact	Significant health impact to *mellifera*
Swarming	Approx. three to six times a year	Generally once a year
Stripes on body	Depends on the subspecies of *Apis cerana*. Japanese *Apis cerana* are black. The *Apis cerana javana* found in Cairns possesses distinctive stripes on its abdomen	Not such distinctive stripes as *cerana*
Main differentiating characteristic	Vein pattern on rear wings different to that of *mellifera*	
Position of bees on brood and honeycomb	*Apis cerana* faces upwards when in a cluster and does the same when working comb	European honey bees face downwards

Approximately 4 million years ago a common ancestor of the Western and Asian honey bees made the Near East and Africa their home and, due to the resulting geographical isolation between the bees in Asia and the Middle East and Africa, they diverged genetically and grew sufficiently apart to become two species.

The lifecycle of *A. cerana* is almost identical to the European honey bee. The table below compares brood development times in days for *Apis cerana* and the European honey bee, *Apis mellifera*.

SPECIES	WORKERS			QUEENS			DRONES		
	Egg	Larva	Pupa	Egg	Larva	Pupa	Egg	Larva	Pupa
Apis cerana	3	5	11	3	4–5.5	6–7.5	3	6	14
Apis mellifera	3	6	12	3	5	5	3	7	14

Apis cerana capped drone cells have a small pin-sized hole in them and European honey bee drone cells do not.

Like the European honey bee, *Apis cerana* make their nests in closed cavities like tree hollows, caves or between rocks or in the walls of houses. The tropical races of both the Western honey bee and *Apis cerana* make less honey than non-tropical races and are also more prone to swarming than their sub-tropical cousins.

Apis cerana have far better grooming behaviour than European honey bees, making them better at brushing off *Varroa* and other parasitic mites. Also, with *Apis cerana* (their natural hosts), *Varroa* infest drone cells, and therefore do not negatively affect the worker population of the colony. In contrast, in *Apis mellifera* colonies *Varroa* infest worker cells (due to these cells being of a similar size to *Apis cerana* drone cells), and negatively affect the worker population.

Across Asia, *Apis cerana* is kept in hives by local beekeepers to produce honey both to consume and to sell to augment income. *Apis cerana* have been domesticated for honey production for many thousands of years and are often preferred over the European honey bee for domestication by poorer native people. This is because the set-up and on-going management costs for European honey bees are significantly higher than for *Apis cerana*. If a European honey bee colony were to die or abscond, replacing the colony would be much more expensive than for an *Apis cerana* colony. A replacement *Apis cerana* colony would be straightforward and almost free to obtain as the beekeeper need often only pay local village children a very small amount of money to find a nearby replacement feral colony.

BUMBLEBEES

Bumblebees are social insects, like honey bees, with many species nesting in small holes made by rodents and other creatures in the ground. In Australia, one species of bumblebee, *Bombus terrestris*, is found, and only in Tasmania. Bumblebees are easily distinguished from honey bees. They are much larger, are more round, have black and yellow bands on their abdomen and have more hair. Bumblebees make a loud buzzing sound when they fly, hence the name bumble, meaning a humming or droning sound, or a buzz. The buff-tailed bumblebee, *Bombus terrestris*, was most likely introduced by accident into Tasmania and was found there in 1992 where it has been successfully living ever since.

There are about 250 species of bumblebees in the world. The majority of species are found in temperate climates in the northern hemisphere, with the centre of bumblebee diversity being Central Asia where most species fly between March and October. In the Southern Hemisphere, in Tasmania and in New Zealand where they are also found, bumblebees fly between September and June.

Bumblebees were deliberately introduced into New Zealand in 1885 to pollinate red clover. Up until that time red clover seeds were imported from England every year at a significant cost to the farmer. Although *Bombus terrestris* is the dominant bumblebee in New Zealand and can be found all over the country, there are three other species found that are less well distributed. These are the large garden bumblebee (*Bombus ruderatus*), the small garden bumblebee (*Bombus hortorum*), and the short-haired bumblebee (*Bombus subterraneus*),

Short-haired bumblebees from New Zealand were being considered as a population source from which the species could be re-introduced into the United Kingdom after they died out there, but the bees were found to be too inbred to be suitable, and a Swedish population of the species was chosen instead.

Bumblebees, like honey bees, rely on collecting nectar and pollen from a variety of floral types to remain healthy

19.6

19.7

19.6: Worker of the bumblebee (*B. terrestris*) with the distinctive white patch at the rear.

19.7: Commercial colonies of bumblebees being used to pollinate tomatoes inside a greenhouse in New Zealand.

and survive. Nectar and pollen are used as sources of carbohydrates and protein to raise brood, while nectar alone is used to provide energy to adult bees for flying or keeping the colony warm.

Bumblebees do not produce honey, but instead store small amounts of nectar. Whereas honey bee colonies produce honey to provide food during the colder winter months, bumblebee colonies die off annually with only newly mated queens overwintering to see the following summer. Due to this fundamental difference in lifecycle, they do not need to produce honey stores. A colony of *Bombus terrestris* typically only contains about 400 workers during the peak nectar flow season in the spring, while a colony of honey bees contains about 50,000 workers.

Bumblebees live in habitats where there are flowering plants such as grassland, woodland edges, roadside verges, gardens and other undisturbed grass areas.

One of the noticeable points about *Bombus terrestris* is the white patch at the bottom of the worker's abdomen. This is absent in the queen, who instead has a yellow band at the end of the abdomen.

Again, like the honey bee, there are three castes of bumblebee — queens, workers and drones. Typical characteristics of the castes of *Bombus terrestris are* as follows.

Queen
- A queen is about 22 to 24 millimetres long.
- She possesses a sting that she will use in defence if provoked.
- The queen can sting multiple times, and does not lose her sting like a honey bee does, and therefore will not die.
- Unlike honey bees, the bumblebee queen will construct comb for her nest, will forage for nectar and pollen during the early stages of nest construction and during times of extreme resource limitation, and will incubate her young.
- The queen has a bald patch on the bottom of her abdomen so that she can efficiently transfer her body heat to brood cells.
- A bumblebee queen will live for one year from autumn to autumn and the old queen will always die before the onset of winter.

Workers
- Workers are about 11 to 17 millimetres long.
- They are able to sting multiple times as they do not lose their stinger during an attack. Bumblebees are usually less aggressive than honey bees.
- The antennae of the worker are shorter than those of the drone and can be used to distinguish between the two castes. A worker has twelve segments on her antennae while a drone has thirteen segments.

- The gender of a bumblebee can also be determined by counting the number of segments on its back: six segments for a worker and seven segments for a drone.
- A worker lives for about eight to twelve weeks.

Drones

- Drones are about 14 to 16 millimetres long.
- They do not possess a stinger.
- The male abdomen is narrower and blunter at the end than that of the worker abdomen.
- There are many more drones during the bumblebee mating season in late summer than at other times of the year.
- Unlike honey bee drones, bumblebee drones do not die during the act of mating, and can therefore mate multiple times if given the opportunity.
- Bumblebee drones forage on flowers just like male solitary bees, and unlike honey bee drones which must be fed in the hive by the female workers.

Bumblebee colonies last for one year and only the new queens survive past autumn. The new queens will have mated in the late summer before going into hibernation for the winter. The new queen will overwinter alone and will wake during a warm day in early spring to emerge from her resting place to search for a location for a new permanent nesting place for her soon to be established colony. The new nest is usually made of moss, grass and leaves in a hole in the ground, often hidden by long grass. In the United Kingdom, some *Bombus terrestris* nests have been found in holes 2 metres deep. A favourite nesting place for *Bombus terrestris* is under garden sheds.

Once the queen has selected the new nest site she will start building the nest by making a single thimble-shaped beeswax cell called a honey pot. The queen uses the honey pot to store honey that she has collected herself from flowers during the early spring. If a queen is seen flying with pollen attached to her hind legs then she has already started to build a nest, since this activity occurs only after she has selected a site for the new nest.

The queen stores pollen in a ball and lays about six eggs at a time in a clump in a depression she has built in the pollen. She then seals the pollen ball with wax and incubates it at between 30° Celsius and 32° Celsius to keep it warm. The queen feeds on honey from the honey pot, which is positioned within her reach so that she can continue to incubate the eggs without needing to move to obtain food.

The eggs hatch and the larvae feed on their bed of pollen. As they grow, the queen opens the wax covering and adds more pollen and nectar. She rests on the brood clump, incubating the larvae to speed their development. The larvae mature and each spins a cocoon of silk in which it pupates and completes development into an adult. After the new workers emerge their empty cocoons are used as storage pots for honey or pollen.

Similar to the European honey bee, the queen lays two kinds of egg: the fertilised worker egg and the unfertilised drone egg. The female worker eggs are fertilised with the sperm that she has stored from her mating flight the previous summer. Unlike the European honey bee, which mates with between ten and twenty drones, buff-tailed bumblebee queens are monandrous, mating with only a single drone.

Typical development times for brood are:

- eggs hatch to larvae after five days
- larvae pupate after a further twelve days
- pupae metamorphise into adult bumblebees after a further fourteen days.

The development time of a typical *Bombus terrestris* bumblebee is about 30 days, depending on the temperature of the brood: a shorter period if the temperature is warmer, and longer if the temperature is cooler. The species of bumblebee will also affect the duration of the brood development cycle.

The first workers to emerge will remain in the nest for about a week before becoming foragers gathering nectar and pollen outside. Once the first set of brood has developed into foragers they take over from the queen the roles of collecting food and feeding the brood. At this time the queen stops venturing outside of the nest and devotes herself to laying and incubating eggs; these she lays in batches, not continuously like a honey bee queen.

Once nectar and pollen become readily available in the spring the *Bombus terrestris* colony grows rapidly to about 400 adult workers. Depending on the supply of food, this size will remain stable until late summer when the old queen will lay drone eggs as well as raise new queens.

The male drones will leave the nest to find a queen and will not return to the colony. Drones sleep outside at night on flowers or in other sheltered places and spend their waking hours patrolling set routes in the hope of coming across a virgin queen. A new virgin queen will leave the nest in search of a mate. The exact sequence of events leading to mating is still unknown in *Bombus terrestris*, but it is thought that queens are attracted to pheromone marks left by male bumblebees at regular intervals along their patrolled route.

Bumblebees are different to honey bees in that during the winter the entire colony dies except for the new queen who goes alone into hibernation. As noted previously, once spring comes the new queen will emerge on a warm day and look for a suitable site at which to build her new nest.

Bumblebees are excellent pollinators and there is a growing demand from some agriculturalists to import these creatures into mainland Australia for the greenhouse pollination of tomatoes. This move, however, is being resisted by others who believe that native solitary bees can perform this role as well, and that to introduce bumblebees into mainland Australia is importing another invasive species. There is a view held by some that bumblebees may preferentially pollinate non-native plant species, facilitating their spread. Furthermore, along with feral honey bees, bumblebees could impact native pollinators by competing for resources and affect cavity-nesting native animals by competing with them for nest hollows.

The bumblebee is able to forage from dawn until dusk and can fly up to 5 kilometres in search of food, pollinating large areas of plants. The temperature at which a bumblebee can fly is much lower than for a honey bee, as low as 5° Celsius. For this reason bumblebees are more effective at pollinating plants in cold or wet regions where honey bees would experience difficulty remaining outdoors.

Many species of bumblebees have longer tongues than honey bees and they can often collect nectar from more inaccessible places in the flower. Also, because bumblebees buzz (or shake) pollinate, that is they grasp the male stamen with their mouth pieces and shake it so that more pollen falls off, they are able to collect pollen from many plant species which honey bees are unable to pollinate.

As an example of buzz pollination, the male stamen of tomatoes only requires a small amount of vibration for pollen to fall off onto the female stigma of the flower. In nature this is performed by the wind and by insects, but in greenhouses where there is no wind and very few insects, bumblebees can be introduced to efficiently perform this role.

Bumblebees are good at pollinating many plants that are often grown in greenhouses, such as tomatoes, cucumbers, zucchinis, capsicum, strawberries, melon, avocados, and kiwi fruit. They readily adapt to confinement in large greenhouses, which honey bees do not, and their effectiveness as a pollinator makes them a useful ally to the horticulturist.

WASPS

Both the European wasp, *Vespula germanica*, and the Common or English wasp, *Vespula vulgaris*, are found in mainland Australia and Tasmania. The two types of wasp are almost identical to look at and the two points of difference between them almost impossible to detect unless the wasps are stationary. Firstly the patterns on the face differ and secondly the European wasp also has two black spots on the top-most yellow band of the abdomen.

As the European wasp is the most prevalent of the two types most of this section will discuss this particular species. In practical terms the European wasp and the English wasp may be regarded as the same and the comments in this section about the European wasp will equally apply to the English wasp. These similar characteristics include lifecycle, nesting, venom, most body markings and aggressiveness.

The European wasp was first found in Tasmania in 1959 having arrived from New Zealand where it had established itself in 1945. By 1978 the European wasp had travelled to mainland Australia and was first found in Melbourne in Victoria.

At present, distribution of the European wasp appears to be restricted to the cool and wet climates of coastal southern Australia and it occurs throughout most of Victoria and Tasmania. In country New South Wales, nests have been located at Coonabarabran while several nests have been recorded in south-east Queensland. In South Australia, the European wasp is well established throughout the hills surrounding Adelaide and within Adelaide itself. In Western Australia, the European wasp has been recorded from Perth to Albany.

About a year before the European wasp reached Tasmania, the English wasp, *Vespula vulgaris*, was found in the eastern suburbs of Melbourne. The English wasp has not been as successful as the European wasp in acclimatising to Australia and has limited spread in the eastern parts of Melbourne and in Gippsland.

Studies in Europe have shown that European wasps are similar to honey bees in that they are found in large communal nests headed by a queen and surrounded by many thousands of female workers. The size of the European worker wasp is between 1.2 centimetres and 1.6 centimetres long. Male drones are laid later in the season and only represent a small fraction of the wasps in the colony.

The entrance hole to a wasps' nest in the ground is typically about 25 to 30 millimetres wide and leads to a tube or tunnel about 1 metre long. The nests are usually built underground and the wasps are able to excavate dirt to enlarge the nest cavity — typically about the size of a football.

Like the honey bee, wasps make cells to raise brood and to store food although cells are made of chewed wood fibres not wax. The colour of the wood fibre comb is greyish, reflecting the colour of the masticated wood fibres.

European wasps are similar to bumblebees in the construction of new colonies, although quite different to the honey bee. Wasp nests are started anew every spring from a single queen that has hibernated alone over the winter months. On a warm day in spring the queen wasp will emerge from her dormant state and will leave her winter's hiding place in search of nectar, honeydew or some other carbohydrate. Honeydew is the sugary secretion of aphids and is collected by wasps, native social bees, and honey bees as a form of food. Once the queen has fed she will look for a permanent home

19.8: The European wasp, *V. germanica*, in Tasmania. Notice the one large and two small black spots on its face, just above its mouth, which can be used to identify this species of wasp in Australia.

19.9: A European wasp. Notice the pairs of black spots on its abdomen, which can also be used to identify this species of wasp in Australia.

19.10: The Common or English wasp. Notice that there are not three spots on the face and that the black 'nose' is longer and more distinct than for a European wasp.

19.11

19.12

19.11: A Common or English wasp, *V. vulgaris*, killing a honey bee.

19.12: Nest of a European wasp. The nest is made of chewed wood and is similar to paper.

19.13: English wasp queen beginning to build a nest and lay eggs.

19.13

in which to build a nest and raise her new colony. Once the new nest site has been selected, the queen will build inside it about 30 brood cells and lay eggs in them. The eggs will turn into grub-like larvae and will develop into adult wasps. These wasps will then take over the role of collecting food, collecting wood fibres to build the comb, and raising the brood, leaving the queen with the sole responsibility of laying eggs. Wasps can often be seen scraping wood fence posts and chewing the wood fibres into a moist paste ready to be used to construct new cells in the nest.

As the summer progresses the colony will grow in size until, in the early autumn, a batch of new queens and male drones are laid, ready for mating later in the autumn. The male drones and the queens will leave the nest in search of mates; the drones will soon die and the mated queens will seek a suitable sheltered location in which to hibernate over the winter. Late in the autumn, larger, slow-flying queens may be seen looking for a suitable place in which to spend the winter months.

Once the queen has selected a home for the winter she will hold on to the structure of the cavity with her mouth; her legs, wings and antennae are all folded back closely against her body. In Europe during late autumn the original queen will die and the nest will slowly disintegrate before the onset of winter as the workers die out.

Workers of the European wasp eat meat as well as liking sweet foods such as nectar, honeydew and soft drinks. Wasps are also a pest of honey bees, robbing honey from hives as well as eating brood and cannibalising adult bees. An attack by wasps on a hive is distinctive in that a wasp, once it has caught an adult bee, will remove the bee's head, abdomen, legs and wings, and return to its own nest carrying only the thorax as this is the part of the bee that contains a lot of nutritious muscle. Workers raise brood on masticated, or chewed, meat protein from insect prey or scavenged meat or carrion.

The foraging range of a wasp rarely exceeds 1 kilometre and is usually about 500 metres. A wasp on a foraging trip typically has a zigzag flight path while it is looking for food, and will often fly close to the ground to look for opportunistic feeds there. On its return journey to the nest the wasp flies in a distinctive straight line and it is possible by looking at the return path of two wasps returning to the same nest to triangulate and make a good estimate of where the nest is located.

Studies in New Zealand have shown that in one year about 4.9 per cent of honey bee colonies were seriously affected by wasps, with 1.9 per cent of the total colonies being completely destroyed. This can be compared with infection by American Foul Brood, which destroyed 0.44 per cent of hives during the same period.

A wasp is more likely to attack humans than is a honey bee. A wasp can repeatedly sting its victim without being killed itself. Studies have shown that the pain inflicted by a wasp sting is the same intensity as a bee sting. Peak aggression of the wasp is during the autumn when colony size is at its maximum and food resources are short.

DIFFERENCES IN COLONY LIFECYCLE BETWEEN EUROPE AND AUSTRALIA

Although the above is true for wasp colonies in Europe, for reasons that are still unknown wasp colonies in warmer parts of both Australia and New Zealand show a very different lifecycle during the winter. In these countries the colony does not die out but survives, retaining up to 100 new queens over the winter. An obvious factor may be the shorter warmer winters experienced in Australia and New Zealand, some two months long compared to five months in much of Europe. With so many resident queens all surviving with the workers over the winter the colony comes out of winter extremely strong and may build up to 100,000 workers and 1 million cells in a nest; this is four times the maximum nest size of any wasp colony found in Europe. In Europe, a typical wasps' nest contains approximately 3000 to 4000 wasps and 12,000 to 13,000 cells.

19.14

19.15

19.14: The Asian hornet, *V. velutina nigrithorax*, is causing significant problems to beekeepers in France.

19.15: Nest of an Asian hornet colony photographed during winter in France.

ASIAN HORNETS

Another pest related to the honey bee, the Asian hornet, *Vespa velutina nigrithorax*, although not found in Australia or New Zealand, is causing significant problems for beekeepers in France since its accidental introduction into that country (perhaps from China) in 2004. The Asian hornet is larger than a European honey bee with queens up to 30 millimetres and workers up to 25 millimetres in length. Asian hornets are easily recognised since their thorax is a velvety black or dark brown with distinctive yellowy-orange stripes. The head of the hornet is black with a yellowy-orange face.

The Asian hornet makes its nest in the open, often hanging from trees, using a papier-mâché material made from chewed wood or plant material. The almost circular nest of the Asian hornet is easy to recognise as it is built in the open and is often about 1 metre in diameter.

The Asian hornet is a very aggressive predator of the European honey bee and can lead to the destruction of the entire colony. Asian hornets will position themselves about 30 centimetres from the entrance to a *mellifera* hive and attack returning honey bees that are carrying pollen. The hornet will then take the honey bee to a nearby tree where it will remove the legs and wings and turn the remaining parts of the bee into a small meatball that it will take back to its own nest as food for larvae.

If the attack is severe, with multiple hornets attacking simultaneously, the effect on a *mellifera* colony can be catastrophic. The attack could, over a short period of time, lead to the complete destruction of the hive or in lesser attacks the *mellifera* colony could be so weakened so as to become vulnerable to other diseases or infestations by pests such as *Varroa* or small hive beetle.

19.16

OTHER RACES OF *APIS MELLIFERA*

The majority of this book is about the European honey bee, *Apis mellifera*, but as mentioned in the section on the biology of the honey bee, other members of the species *Apis mellifera* can also be found throughout Europe, most of Africa and much of the Middle East. Any of the ten distinct races of *Apis mellifera* found in Africa would be classified as unwanted pests if they were to be found in Australia.

19.16: *Apis mellifera scutellata* is native to east and parts of southern Africa. The accidental introduction of *A. mellifera scutellata* into Brazil and the subsequent interbreeding of this race with members of the European honey bee races led to the emergence of Africanized bees and their subsequent advance northwards to southern areas of the United States.

Some African races of *Apis mellifera* are very aggressive; for example, *Apis mellifera adansonii* from Ghana or *Apis mellifera scutellata* from central and southern east Africa. Although any of the ten races of honey bee found in Africa can technically be called African honey bees, it is for *Apis mellifera scutellata* that this name is most commonly used.

The African honey bee, *Apis mellifera scutellata* was introduced to Brazil in 1956. It escaped from a laboratory and slowly made its way north, eventually reaching the southern parts of the United States in 1990. Even though *Apis mellifera scutellata* bees had crossed the Amazon basin and traversed the entire Central American isthmus spreading through Mexico, there was very little interbreeding with the small numbers

A. mellifera scutellata

19.17

19.18

19.19

19.20

19.17: Distribution of A. mellifera scutellata in Africa.

19.18: Africanized honey bee (top left), next to a European honey bee (bottom right).

19.19: Cape honey bee, Apis mellifera capensis, on a flower.

19.20: A. mellifera capensis queen and her retinue.

of European honey bees that were present in these areas. Most African bees, by their nature, are usually aggressive and *Apis mellifera scutellata* is no exception. The limited cross breeding of *Apis mellifera scutellata* with the existing varieties of *Apis mellifera* has given rise to the name Africanized bees, or, more popularly, Killer bees, to distinguish them from the purer races of African bees found in Africa. If either *Apis mellifera scutellata, Apis mellifera adansonii* or most of the other races of African honey bees were to enter Australia, it is likely that the result would also be some very aggressive bees in Australia that would pose a significant threat to public health. Even today in Ghana *Apis mellifera adansonii* is so aggressive that beekeepers only work these bees at night and change into their protective clothing before getting out of their vehicles.

The commonly called Cape bee, *Apis mellifera capensis*, native to the Cape region of South Africa, would present a very different problem to beekeepers in Australia if it were to be found here. In southern Africa, around the Cape, *Apis mellifera capensis* beekeepers were successfully keeping this race of honey bee for honey production. In 1990, in order to further exploit this race they took colonies of the Cape bee north to areas that were home to the African bee, *Apis mellifera scutellata*. This created a unique problem because, unlike other races of *Apis mellifera* where workers can only lay unfertilised drone eggs, *Apis mellifera capensis* workers can lay fully fertilised worker eggs that contain both the mother's DNA and the father's DNA, resulting in a viable worker being born without the *Apis mellifera capensis* worker first mating with a drone. Once the Cape bee had been taken north into territory that was home to the African bee, Cape bee workers drifted into the hives of African bees where they promptly started to lay viable worker Cape bee brood. This led to a situation where inside an *Apis mellifera scutellata* colony there was both an *Apis mellifera scutellata* queen laying eggs and many Cape bee, *Apis mellifera capensis*, workers laying eggs. In this situation the number of Cape bees inside the colony soon overtakes the number of African bees and the original colony ultimately absconds or collapses. This situation is currently causing significant disruption to beekeeping in the northern parts of South Africa and there does not appear to be a solution to the problem.

OTHER SPECIES OF ASIAN BEES

There are believed to be twelve species of honey bee globally.

Cavity nesting bees

1. *Apis mellifera* — European or Western honey bee
2. *Apis cerana* — Asian honey bee
3. *Apis koschevnikovi*
4. *Apis nuluensis*
5. *Apis nigrocincta*
6. *Apis indica*

Dwarf honey bees

7. *Apis andreniformis*
8. *Apis florea*

Giant honey bees

9. *Apis dorsata*
10. *Apis laboriosa*
11. *Apis binghami*
12. *Apis breviligula*

As mentioned previously, *Apis cerana* is commonly called 'the Asian honey bee', while all of the above species collectively, excluding *Apis mellifera*, are called Asian honey bees.

19.21

19.22

19.21: *A. florea* worker.

19.22: *A. florea* make their small nests in the open under small branches.

DWARF SPECIES OF ASIAN HONEY BEES

I mentioned earlier *Apis cerana,* the Asian honey bee, which entered Australia through Cairns. The other ten species of Asian bee could also successfully establish themselves in Australia were they to land on our shores. One example would be the dwarf honey bee, *Apis florea.* The smaller Asian bees, such as *Apis florea,* differ from both *Apis cerana* and the European honey bee in that they form nests in the open, hanging from small branches often deep in bushes where they remain hidden. As a defensive strategy the smaller Asian honey bees abscond very quickly, often at the slightest physical disturbance.

The dwarf honey bee, *Apis florea,* constructs a single comb nest in the open and does not collect or store much honey. The bee is gentle and, due to its small size, is unlikely to be recognised as a honey bee when visiting a flower. Due to its small size, gentle behaviour, and shyness, any incursion into Australia is likely to remain undetected for some time, allowing the bee to become well established before eradication can be attempted.

Apis florea has spread in recent years from its original home in south-east Asia and can now be found widely in east Africa, Israel, Jordan, Oman, and across the Saudi Arabian peninsula. It is not clear how much of this range expansion is due to human relocation, and how much is due to the natural spread of the species. Since the 1980s, *Apis florea* has been spreading at a rapid pace throughout Sudan after being accidentally introduced into Khartoum, and is now found throughout 90 per cent of the country.

GIANT SPECIES OF ASIAN HONEY BEES

The four giant honey bee species found in Asia are *Apis dorsata*, *Apis breviligula*, *Apis binghami* and *Apis laboriosa*. Until recently, *Apis breviligula* and *Apis binghami* were considered to be subspecies of *Apis dorsata*, but recent genetic studies suggest that they are a distinct species. The four giant honey bee species make their nests in the open, typically under cliffs, under large branches of trees or under the roofs of large buildings such as minarets. *Apis dorsata* and *Apis laboriosa* often nest in large aggregations of tens to hundreds of colonies, while *Apis breviligula* and *Apis binghami* tend to make single solitary nests. The giant honey bees, although successfully hunted for honey across Asia, India and Nepal, cannot be kept in hives and the collection of their honey represents a significant personal safety risk for the honey hunters who depend on them for their livelihood. Although this safety risk is partly due to the aggressiveness of the giant bees, part of the risk is also due to the heights that honey hunters need to climb in order to harvest comb. The honey produced by the various giant honey bees is considered by locals to be far superior in taste to that of other honey bee species, making the reward to be gained from the harvest worth the risks involved.

19.23: A permanent *Apis dorsata* nest in Asia. *Apis dorsata* make their nests in the open on trees or buildings.

19.24: *A. laboriosa* worker taken in India in the Himalayas. *A. laboriosa* are found at higher elevations in the Himalayas while *A. dorsata* are much more widespread and are found at lower altitudes.

19.25: *A. dorsata* from the Philippines.

19.23

19.24

19.25

20.1

20.2

20.

Flow Hive

INTRODUCTION

Beekeeping is a traditional activity that changed little over the last 150 years until Stuart and Cedar Anderson came along in 2015 with an innovative design for frames that changed beekeeping and extracting. Flow hives generated a lot of interest among existing and retired beekeepers as well as those who may not normally have taken up beekeeping. Hobby beekeepers and many of those interested in beekeeping were keen to try Flow as it makes harvesting honey easier. Extracting honey was certainly made easier, although harvesting honey is only a small part of beekeeping. Making sure the colony is strong, has a good queen, is free of pests and diseases, and is less likely to swarm, still takes up most of a beekeeper's time. The surge in hobby beekeeping in recent years owes much to Flow.

Starting as a beekeeper with a Flow Hive is the same as starting with a Langstroth hive. A brood box is needed together with bees. This part is the same as described in chapters 5 and 6. As the colony gets stronger and needs to expand, a Flow super that contains Flow frames is placed above the brood box. Worker bees use Flow frames to store honey, which, when ripe, can be extracted with minimal disturbance to the colony.

Flow is designed so that extraction takes place at the hive without the need to open the super. This is very different to traditional Langstroth hives where the super is opened, honey frames are removed, taken to the garage for extraction, and then returned to the hive, which requires opening it once again. Using the Flow frames eliminates this time-consuming, disruptive and messy process of honey extraction. Since there is little disturbance to the bees, the likelihood of being stung may be greatly reduced.

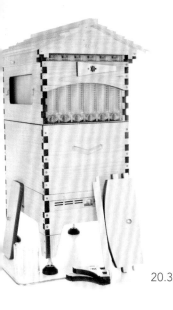

20.1/20.2: Flow Hive Classic and Flow Hive 2 make attractive additions to a garden.

20.3: Flow Hive 2.

20.4: Flow Hive Hybrid.

20.3

20.4

FLOW DESIGN

Three models of Flow Hive are available: Flow Hive 2, Flow Hive Classic and Flow Hive Hybrid. Flow Hive 2 was introduced in 2018 and added features that make beekeeping easier compared with Flow Hive Classic. Both designs contain Flow frames only in the super.

When you buy a Flow Hive it arrives as a kit, like an IKEA kit, and includes detailed instructions and the tools needed for assembly. The Flow frames themselves arrive fully assembled. The people at Flow have placed many excellent videos on their website (https://www.honeyflow.com) and on YouTube that clearly show how to assemble a Flow Hive, keep bees and extract honey, together with lots of other useful information. Many other Flow users have also uploaded material to YouTube, so there are plenty of videos to watch. Google 'Flow Hive' and take your pick.

20.5

20.5: Flow Hive Classic.

A Flow Hive can be bought either as a complete hive, including Flow frames, Flow super, base, brood box with wooden Langstroth frames, excluder, top cover and roof, or just as a single Flow super with Flow frames to put on your existing hive. Individual frames can also be bought. Flow supers and brood boxes are the same size as traditional eight- or ten-frame Langstroth hives, so a Langstroth base, brood box and Flow super

20.6: Roof of a Flow Hive.

20.7: A top cover used instead of a hive mat.

20.8: Rear of Flow Hive super showing removable panels.

fit on top of each other without modification. If you plan only to buy a Flow super, make sure you get the eight-frame or ten-frame sized super to fit your existing hive.

Starting from the top, the hive includes a sloping roof. Similar to a telescopic lid, the roof overhangs the super on all sides and there are thumb screws on Flow Hive 2 to attach it to the super so it does not blow off. Underneath the roof, between the roof and the super, goes the top board. Top boards are widely used in North America, although most Australian beekeepers use a hive mat. In my experience, top boards are more practical than hive mats and easier to use. The top board has a hole that allows air to circulate and can be used for feeding syrup.

Next comes the super that holds the Flow frames and has a different back end to a Langstroth super. Because Flow is designed so that honey can be extracted directly from the super without removing any frames, the back provides access to frames via a removable panel, which is kept in place by a swivelling piece of wood. Removing the

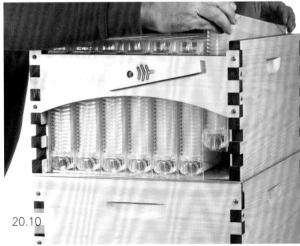

panel allows you to attach harvesting tubes to the bottom of the frames through which honey can be extracted. At the top of the rear of the super is another removable strip of wood through which you insert a Flow key into each frame.

The side window on the super and the rear view between the Flow frames can be used to check how much honey is present and if it needs extracting. The windows can also be used to watch the bees as they go about their daily tasks.

Underneath the super goes the queen excluder and underneath that the brood box. These are identical to those used in Langstroth hives.

Flow Hive includes a queen excluder and I recommend you use it to ensure the queen stays in the brood box and does not lay eggs in Flow frames (see p. 56). Cells in Flow frames are slightly deeper than cells a worker would make on wax foundation; this discourages queens from laying eggs in Flow frames, although it sometimes happens.

20.9: Panels can easily be removed and replaced.

20.10: Flow super with frames in place.

20.11: Flow Hive has a window on the side that can be used both to check the colony as well as teach children about bees. Flow Hive 2 has windows on both sides of the super.

The Flow base is different to those used in Australia, although North American beekeepers will be familiar with the design. The base includes a tray for monitoring *Varroa*, which is now in Australia, and provides ventilation for hot or damp conditions. The design of the base can also help control Small Hive Beetle and Wax Moth, since

20.12

20.13

20.14

20.15

20.12: Aerated base used in Flow Hive Classic showing removable core flute insert.

20.13: Although different from typical Australian bases, the design is standard in North America.

20.14: Assembled base for Flow Hive 2 showing removable tray that can be used for pest management.

20.15: Flow Hive 2 base showing optional adjustable legs.

beekeepers have easy access to the tray that can hold Small Hive Beetle traps or moth larvae that can be easily removed. Flow Hive Classic has a different base to Flow Hive, although both include a removable tray.

FRAMES FOR BROOD BOX

Flow includes traditional foundationless Langstroth frames in the brood box. At the top of the frame is a thin strip that workers use to start building comb. Natural beekeepers have always used this technique both in Langstroth hives and in top bar hives (p. 73).

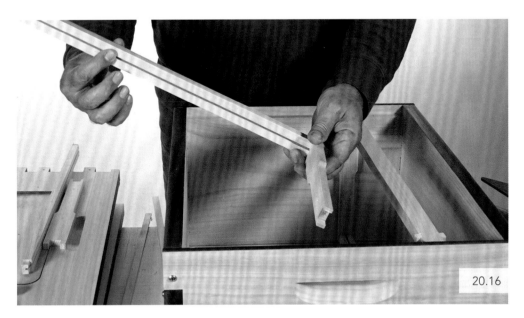

20.16

20.17

20.16: Frames for the brood box need to be assembled before use. Flow frames for the brood box do not use wax or plastic foundation.

20.17: Inserting the strip of wood into the top bar from which worker bees will draw comb.

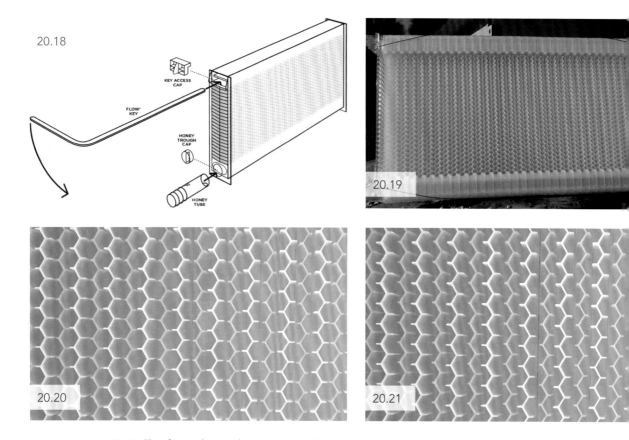

20.18: Flow frame showing key access cap, honey trough cap, honey tube and Flow key.

20.19: Flow frame comes assembled with the hive kit, or it can be bought separately.

20.20: Cells of Flow frame are closed for bees to store honey.

20.21: Cells are opened during extraction, allowing honey to escape. The Flow key is used to open the cells in frames.

FLOW FRAMES FOR SUPER

Flow frames are designed so that honey can be extracted from them without removal from the super. When the key is inserted into the top and turned, the cells in the frame move apart, causing honey to flow to the bottom. A channel at the bottom collects the honey and it flows out of a tube at the back into a container.

WHICH FLOW HIVE MODEL TO CHOOSE?

The design and operation of Flow Hive 2 and Flow Hive Classic are almost identical, although Flow Hive 2 is easier to use as some useful features have been added. Visit the Flow Hive website for a description of the improvements.

If you do not want to buy a complete Flow Hive, just add a Flow super to your existing Langstroth hive. Although you can buy individual Flow frames, the design of the Flow super is patented, which restricts woodworkers from making their own super based on the Flow design. If you decide to add a Flow Classic super to your Langstroth hive and use a telescopic lid (p. 60), the lid may no longer fit, since the knobs on the top strip of the super will stop the lid from hanging over the edge. To overcome this, place the Flow super above the queen excluder and then place another Langstroth ideal or full-size super above the Flow super. The telescopic lid now fits as intended. Bees prefer waxed comb or foundation frames over a plastic frame, so putting an extra, regular super on at a time of limited nectar flow may mean that the Flow frames will not be filled.

An eight-frame Langstroth super holds six Flow frames, while a ten-frame Langstroth holds seven Flow frames.

SETTING UP A FLOW HIVE

Setting up a Flow is the same as starting with any kind of hive. Find a level area of ground to place the base, visually or using a spirit level, and make sure that the base slopes backwards between two degrees and five degrees, allowing honey to flow to the back of the frame for extraction. A traditional Langstroth hive tilts forward, allowing condensation to escape; the Flow hive aerated base does this job instead. Place the brood box on the base, add bees, then place the inner cover and roof.

When the brood box is full of bees, add an excluder and Flow super on top of the brood box, then replace the inner cover and roof. Inner covers are not widely used in Australia where hive mats are common, although they are almost exclusively used in the US. Although slightly more expensive, I find that inner covers are more practical and last the life of the hive, while hive mats need to be replaced periodically.

20.22: Two spirit levels are built into the hive to make levelling easier.

MIXING FRAME TYPES IN A FLOW SUPER

You may want to mix Flow and Langstroth frames in the same super, placing them next to each other. Flow frames should be in the centre as bees prefer storing honey on Langstroth frames and by placing Flow frames in the centre they are filled with honey first. A combination of Flow and Langstroth frames in a super can be purchased from Flow and is called Hybrid. You can also buy your own Langstroth frames and use them in combination with Flow frames.

EXTRACTING

Extracting with Flow, although simpler than conventional extraction methods, still requires care and attention. Frames should be extracted when they are about 80 per cent full of capped honey. Since extraction is intended to be done without opening the hive, you need some experience to determine which frames are full and ripe. In the first year, beginners should regularly open the hive to inspect both the brood and the super (chapters 5, 6). The brood is inspected for colony health and the Flow frames are inspected to determine the ripeness and amount of honey present. After a season a Flow owner will be able to gauge the ripeness and amount of honey just using the weight of the hive (p. 138, picture 8.10) and the views through the rear and side windows.

If an extracted frame is not 80 per cent capped there is a risk the honey will be too watery and so will ferment. Open the side viewing panel and, if the end frame is 80 per cent capped, probably all frames are capped since end frames are usually the last to be filled. If the end frame is not capped you need to open the hive, remove a central frame, and check how much of it is capped. If only one frame is capped you can extract that, otherwise you might need to wait a week and check again. Another way to check is to slightly open one of the central Flow frames. A small amount of honey will escape that can be tested for water content using a refractometer (p. 170). If the water content is below 20 per cent it is safe to extract.

Follow these steps when extracting:

- Before extracting you need to remove both the wooden strip at the top of the super as well as the large panel. Remove the plug at the bottom of the frame and insert the clear plastic tube into the hole at the bottom. Underneath the end of the tube place a large jar to collect honey.
- Next, insert the long Flow key into the top hole, making sure to insert it under the bar across the centre of the hole, into the bottom slot. To open the frame, insert the key below the bar; to close the frame, insert the key above the bar (20.26). To make opening the frame easier you can insert and turn the key progressively a few centimetres at a time. This means you will be operating a much smaller segment of the frame. While slower, this is much easier to perform; it also insures against the possibility of too much honey flooding into the bottom channel at once. Turn the rod slowly either clockwise or anti-clockwise, it does not matter which. If you have not already bought a Flow Hive, when you place your order buy an additional Flow key since opening a whole frame is much easier if two keys are used, one being turned clockwise, the other anticlockwise.

- Once the frame has been opened, honey will start running into the channel, and from there into the jar.
- After the honey has stopped running, insert the key above the bar across the top and turn it 90 degrees. This will realign the cells to their hexagonal shape. I leave the key in this position for a minute or two to ensure the cells have fully returned to the closed position. Replace the plug, which was designed so that it cannot be inserted if the frame has not been closed.
- Remove the plastic tube and replace the stopper. There will still be a small amount of honey in the channel at the bottom of the frame. Frames have a small 'Leak Back Gap' in the channel so that residue honey will escape into the brood box where bees will gather and store it. Worker bees will now clean and repair the frame and return to storing honey in it.
- Repeat this for each frame that needs extracting and then return both the top strip and back cover to their original positions. If there is more than one frame to extract, you could extract more than one frame at once. One Flow frame gives about 3kg of honey, about the same as a Langstroth frame.

20.23: Opening the back of the Flow super, ready for extracting.

20.24: Removing the honey trough cap. If the cap is tight use pliers to gently unscrew the cap.

20.25: Inserting the honey tube through which honey will be removed.

20.23

20.24

20.25

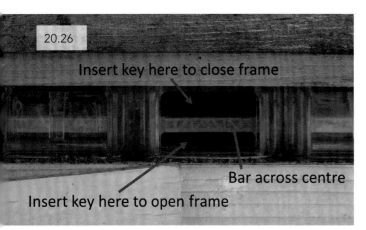

20.26

Insert key here to close frame

Bar across centre

Insert key here to open frame

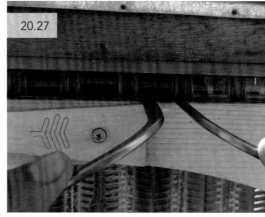

20.27

20.26: Flow key access hole at top of frame. To open the cells in the frame, the key is inserted through the bottom gap and rotated. At the end of extraction, the key is inserted into the top gap.

20.27: The use of two keys makes opening the frame easier. If rotating the keys is difficult, insert the keys a few centimetres at a time, opening the cells progressively.

20.28: Honey flowing from a honey tube.

20.29: Honey flowing into jar during extraction.

20.28

20.29

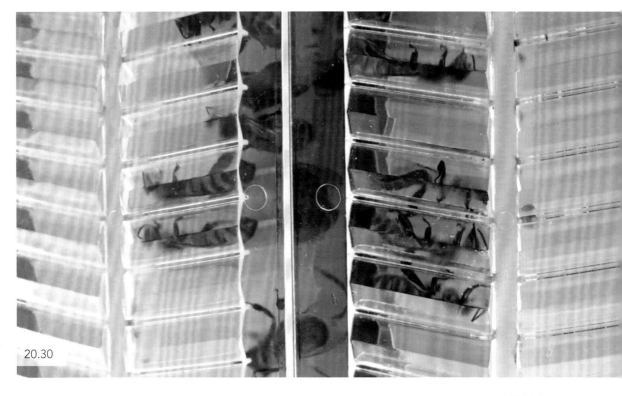

20.30

TIPS WHEN KEEPING BEES

Getting honey to flow may be impossible if the honey comes from canola, tea tree, or Paterson's curse and has crystallised. Crystallised honey in a frame is annoying no matter what system of extraction you use. There are a number of ways to deal with this: leave it for the bees, put the super underneath the brood box with another empty super on top, score the surface of the frame and place it upside down in the hive. You may need to do this for all frames, including Langstroth frames. If this is the case, turn the whole Flow super upside down and leave it for a week after scoring the surface of all frames. If only one or two frames contain thick or crystallised honey, place an empty super under the Flow super but above the queen excluder and leave the upturned frames to be cleaned in it for a week. This is the same as described on pp. 138 and 139. The advantage of this method is that workers will remove the crystallised honey, mix it with liquid honey and store in new cells. The resulting honey should remain liquid and easier to extract.

Honey can be thick and slow to move because of the temperature and/or the type of nectar it comes from. One of the advantages of Flow is that the extraction can be held over several hours or even overnight with no extra effort from the beekeeper. Thus, thicker honey may flow to the jar over time. Remember that the honey in the Flow frames will be warm from the brood below, even in cold weather.

20.30: Flow frames that have been extracted and are being cleaned by worker bees ready to store more honey.

20.31

20.32

20.31: Honey flowing into a jar. If bees around the jar are a problem, the jar can be covered with a plastic bag (shown on right) to keep them away from the honey.

20.32: Different coloured honeys collected from the same hive during extraction. Since frames are filled progressively, generally from the centre frame outwards, as different types of plants produce nectar, different flavours and colours of honey may be collected in different frames.

When extracting, if care is not taken bees will collect around the jar into which honey is flowing. The way round this is to use a plastic tube and plastic bag between the Flow tap and the bottle, or to cover the tap and jar with a net or plastic bag.

When extracting a full super you may be tempted to extract a few frames at once into the same container. Bees generally fill one frame at a time with honey, so frames next to each other may contain honey with a different taste and colour as different flowers come into bloom. Try to extract each frame into a different container. If two jars hold different types of honey, you will benefit by tasting different types of honey.

Bees sometimes will not store honey in a Flow frame. Try rubbing wax across the surface of the Flow frame; workers often start storing honey once this has been done. If you have a hybrid super containing both Flow and traditional frames, bees prefer building cells on wax foundation and storing honey in them. When the Langstroth frames are full, workers will store honey in Flow frames. Another method to familiarise the bees with the Flow frames is to place the Flow super on the base underneath the brood box, with the excluder in between. The bees have to crawl over the Flow frames to get to the brood from the entrance. After a week of this you can return the super to above the brood box.

FEEDING BEES FROM TOP COVER

The top cover has a hole in the centre. In addition to providing ventilation, a common use for this hole is to feed bees syrup, sugar crystals or pollen. Place a jar of syrup over the hole, under the roof, for the bees to eat. This is similar to the method described on p. 122 and is commonly used in the US.

MORE THAN ONE SUPER

If you have a lot of honey coming in or live distant from your hive and are unable to check it regularly, you can use two Flow supers, or one Flow super and one containing Langstroth frames. I would put the Flow super immediately above the brood box to encourage the workers to fill Flow frames first.

PACKING DOWN FOR WINTER

Towards the end of autumn, as colony numbers decrease and nectar is in short supply, it is time to pack down Flow for the winter. If any frames are full of honey, they can be extracted. If frames are only half full you can let the bees clean out the honey, wash them in cold water and store them.

Another way to clean a frame is to leave it in warm, soapy water. The honey will dissolve over time, then rinse the frame with cold water.

20.33

20.33 The hole in the top board can be used to feed syrup to the colony.

DISADVANTAGES OF A FLOW HIVE

A Flow Hive is expensive compared to a Langstroth hive. You need to remember though that if you buy a Langstroth hive you also need access to an extractor, and these are expensive and take up room. Extracting from a Flow Hive is simpler, so Flow beekeepers are not faced with the same work and mess that comes with extracting from a Langstroth hive.

Flow Hives make harvesting honey much easier and quicker than conventional methods. However, work and diligence are still required to care for the health of the bees in the hive.

If the hive is not set up as described on the Flow website, honey is more exposed than with a Langstroth hive and may attract robber bees to the hive (pp. 133 to 135).

Appendix:
Sizes of frames and boxes

Initially the choice of hive size offered to the new beekeeper is between a box that will hold eight frames, that is, an eight-frame box and one that will hold ten-frames, that is, a ten-frame box. More rarely there are beekeepers who use twelve-frame boxes, but these are no longer commonly seen or available from suppliers.

In Tasmania, Victoria and South Australia the majority of beekeepers use eight-frame hives, while those in New South Wales, Queensland, Western Australia, and the Territories favour ten-frame hives. I do not know why this is, but probably the answer can be found in the history of beekeeping and the culture that has grown up in the particular state. It is probably the same rationale as the initial choice of different gauge railway lines across parts of Australia.

Whether you decide to use eight-frame or ten-frame hives you should, for practical reasons, stay with your initial choice and use the same size box for all future hives. Otherwise you will find that having a mix of both eight-frame and ten-frame hives will in the long run cause frustration when you want to rearrange the boxes in your apiary, as the various hive parts will not be interchangeable. I believe that if you have one eight-frame hive and one ten-frame hive, it is worth moving all of the bees, say, from the ten-frame hive into a new eight-frame hive and discarding the empty ten-frame hive. Either that or sell your ten-frame hive and use the money to replace it with a new eight-frame hive. This will save you a lot of inconvenience later. Having said that, of course, if you intend keeping a number of hives you may more easily use both sizes as you will have sufficient of each to manipulate boxes as you please.

I should note here that not all hive and frame manufacturers use the same dimensions for the hive parts and I have sometimes experienced situations where a ten-frame box from one supplier will hold nine frames rather than the expected ten. This fact is not always readily apparent when you purchase a box flat packed and is not generally much of a problem as it is common practice to put nine frames into a ten-frame super or seven frames into an eight-frame super in order to allow the bees to draw the comb out more.

Another point to keep in mind when selecting the size of box is that an eight-frame box holds eight frames with room separating the frames that is usually sufficient for removing the frames without difficulty. The ten-frame box fits ten frames snugly, which can make separating the frames for removal difficult, particularly after winter when the frames have swelled due to moisture. When I use a ten-frame box my preference is to only insert nine frames for ease of removal, particularly when overwintering.

Why would you choose one size box over another? A lot depends on your age and the state of your back. A ten-frame super full of honey will weigh around 40 kilograms; on the other hand a ten-frame super offers you either one or two more frames of honey and a ten-frame brood box offers a queen extra laying space. For me the weight of the respective boxes is the deciding factor.

What is called a 'full-depth' box containing a 'full-depth' frame is a standard used across Australia and can refer to either an eight- or ten-frame hive box. 'Full depth' refers to the depth (measured top to bottom) of the respective box and differentiates these from more shallow-depth boxes that are also available and used by beekeepers. Once you have made your choice of whether to use either an eight-frame or a ten-frame hive box, you will next have to consider the depth of the box. Your supplier or other beekeepers will no doubt offer advice, with most suggesting a full-depth box to sit on the hive base as a brood box. Which depth of box you will be recommended to use as a honey super will again largely depend on which state of Australia you happen to reside in.

In Australia there are currently six different depths of box in use. These are:

- Jumbo or Deep: 280mm deep – designed by Dadant in the United States for use as a brood chamber
- Full-depth: 240mm deep
- WSP: 192mm deep
- Manley: 168mm deep
- Ideal: 144mm deep
- Half: 125mm deep

Apart from the depth, the dimensions of the remaining sides of boxes are:

- eight-frame: 51cm x 35cm
- ten-frame: 51cm x 41cm

The box depths listed above are approximate because as previously noted not all hive manufacturers use exactly the same dimensions. They will be sufficiently similar, though, so that a full-size frame from one supplier will match the depth of a full-size frame from another supplier. Occasionally, though, a ten-frame box will only hold nine frames.

Professional beekeepers in the northern states of New South Wales and Queensland commonly use the shallower Manley or a WSP box, while in the southern states of Victoria and South Australia the WSP is more commonly used. Professional beekeepers often use the shallower WSP or Manley box for supers since back injury is the major cause of disability amongst beekeepers and choosing a lighter box helps alleviate ongoing suffering.

The shallowest depth box available is the Half-depth, although it is the slightly deeper Ideal depth that is most frequently used by hobbyists who want a lightweight box that can easily be

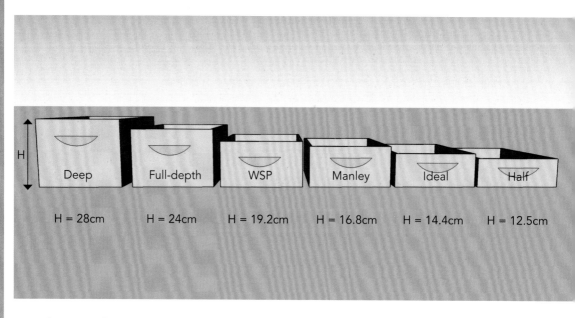

Deep	Full-depth	WSP	Manley	Ideal	Half
H = 28cm	H = 24cm	H = 19.2cm	H = 16.8cm	H = 14.4cm	H = 12.5cm

A.1: Relative size of the different designs of hive boxes. The height of the boxes is given in centimetres.

lifted and carried without back strain. I believe that a major part of the reason for the choice of the Ideal box is due to the fact that many hobbyists are simply not aware of the existence of WSP or Manley boxes. This lack of awareness is often due to a limited range offered by suppliers, either because of the culture of the state, difficulty in obtaining supplies or simple lack of demand by customers.

As a general rule, if you buy an existing hive and you suspect that the hive is home-made, check that the dimensions correspond to one of the standard depths listed above and are designed to hold either eight or ten frames. Many home-made hives are made to apparently random dimensions and finding additional suitable boxes or frames later may be impossible. Many hobby beekeepers with plenty of spare time make their own hive boxes, but usually buy standard frames to fit into them. If you are offered non-standard hives and frames by a seller, I would gently turn down the offer and look elsewhere for standard hives.

The typical beehive consists of one, two, three or more boxes called brood boxes or supers sitting on top of each other. Calling the bottom box a 'brood box' and the boxes on top of this 'supers' is confusing to many new beekeepers as they are in fact the same design of box with the same number of frames. The label 'brood box' is generally adopted for the bottom box, as it holds the brood.

The brood box at the bottom of the hive requires the largest depth box to enable the queen to lay as many eggs as possible during the spring honey flow. This is important since if I am to obtain the maximum number of workers in my hive to collect nectar, the queen needs to be able to lay the maximum number of eggs. To provide sufficient space for the queen to lay eggs I usually use two full-depth brood boxes unless I am building up a small colony from a swarm or a split. The much larger Jumbo or Deep brood box is better at supporting a large number of brood than a single full-depth brood box, but the large Jumbo box is not readily available for purchase in Australia. Hence two full-depth boxes stacked on top of each other provide similar space for the queen.

JUMBO BOXES

Jumbo boxes are never used as supers due to their larger depth and their weight if they were to contain full frames of stored honey. Their use is restricted to housing brood. Since brood boxes are seldom moved by hobbyists, and a professional apiarist would probably use a mechanical lifter to move them, their use as a brood box is seldom a concern.

FULL-DEPTH BOXES

The full-depth box is the standard depth across Australia for both eight-frame and ten-frame hives and is readily available at all beekeeping supply stores.

WSP BOXES

The next box in depth is the WSP, named after its inventor W.S. Pender of New South Wales. These are available in both ten- and eight- frame versions.

As previously mentioned, a full-depth ten-frame super full of honey weighs about 40 kilograms and many beekeepers have strained their back from manually lifting them. The WSP, in a ten-frame size, has less depth and is therefore lighter to lift. The fact that it has less depth and therefore laying space means that the WSP box is seldom used as a brood box and is generally confined to use as a super. Ten-frame WSP boxes are widely used in New South Wales and Queensland although they are not often used in Victoria where the lighter, full-depth eight-frame super is more common.

MANLEY BOXES

Named after Robert Manley (1888– 1978), an English manufacturer of beekeeping equipment, this size box is frequently used in both Queensland and New South Wales and is also the most popular size of box used in the United States.

IDEAL BOXES

The most common small-depth box used by hobbyists is the Ideal. The Ideal is just over half the depth of a full super and I find that many hobby beekeepers who want a smaller-depth super to reduce lifting weight choose this depth box. Personally, the disadvantage I find of an Ideal super is that the wooden sides and the top and bottom bars of the frames take up much of the space inside the box, leaving little space for bees to store honey. Another disadvantage

I find when using smaller-depth supers is that I need three Ideal supers to store the same amount of honey as a single full-depth super. Ideal frames are also harder to make than full-depth frames due to the difficulty of tightening the supporting wire in the shallow frame. Ideal boxes are used extensively in Tasmania for both brood and honey supers.

Extracting can also be more difficult from an Ideal frame, particularly when using an extractor with baskets designed to hold full-depth frames. On the plus side, Ideal frames without wire but with a small insert of starter comb are 'ideal' for drawing comb honey. Despite these perceived difficulties on my part many beekeepers choose to use this depth of box throughout their hive set up.

HALF-DEPTH BOXES

Similar in depth to the Ideal, the half-depth box has the same advantages as the WSP or Ideal in reducing the weight of the super. Although I have never seen a half-depth box used in Victoria they are used in other parts of Australia.

My own preference when selecting a lighter super in which to store honey is to use a WSP, as I believe that this depth of box is the best compromise between the amount of honey stored and the ease of lifting the super. My own experience is that it is preferable to keep the depth of both the brood boxes and the supers the same. This is because when managing hives, frames from the brood box are often swapped with frames from the super and having different depth boxes makes hive management more difficult.

Glossary

ACUTE BEE PARALYSIS VIRUS (ABV): A viral infection of the honey bee that causes paralysis of adult bees.

ABSCOND: This occurs when a colony of bees suddenly leaves a hive with few or no bees remaining behind. Absconding should not be confused with swarming; it is often caused by problems such as such as poor temperature/ventilation within the hive, mite, pest, lack of food, or other intolerable problems.

AMERICAN FOUL BROOD (AFB): A lethal infection of honey bees with the bacterium *Paenibacillus larvae*.

AFRICANIZED HONEY BEE: A term used to describe the South African honey bee *Apis mellifera scutellata* or its hybrids, now found in both North and South America. Africanized honey bees are known for their volatile nature. The term Africanized honey bee is used to differentiate the type of bee found in the Americas from African honey bees, which although of the same race are only found in Africa.

AFTER-SWARM: A swarm which leaves a colony with a virgin queen, shortly after the first (or prime) swarm has departed. The first after-swarm is also referred to as a secondary swarm while an after-swarm that leaves after the secondary swarm is also called a tertiary swarm.

ASIAN HONEY BEE: One of the twelve species of honey bee found globally, the Asian honey bee (*Apis cerana*) is found across large areas of Asia as well as parts of Queensland.

ASIAN BEES: Of the twelve currently identified species of honey bees, eleven are only found in Asia and these are called by the generic name Asian bees. The twelfth type of bee, the European honey bee, *Apis mellifera*, before being transported globally by humans, was only found in Africa, Europe and parts of the Middle East and is thus not an Asian bee. Note that Asian bee is a generic term for eleven species of honey bee found in Asia, while the term The Asian honey bee, AHB, usually applies to the single species, *Apis cerana*.

BAIT HIVE: An empty hive left to attract swarms of bees. The term sometimes refers to hives that have been left near ports to attract swarms that have reached Australia by ship and have eluded capture by quarantine. A bait hive may use the Nasonov pheromone to attract swarms.

BEE SPACE: A space between two frames that is big enough to allow two bees to pass each other while working on opposite frames. A bee space is about 9.5mm wide, which is too small to encourage comb building, and too large to induce propolising activities.

BLACK BEES: *Apis mellifera mellifera*, sometimes called German bees.

BRACE COMB: Pieces of seemingly random comb that connect hive parts together, such as between two frames or between an end frame and the wall of a hive. Brace Comb is a form of burr comb.

BROOD: A general term to describe young bees that have not yet emerged from their cells as adults. Brood can be workers, drones or queens. The four stages of brood are egg, larva, pre-pupa, and pupa.

BROOD CHAMBER: The area of the hive where the queen lays eggs and brood are reared; usually the lowermost parts of the hive.

CAPPED BROOD: Brood in the pre-pupal or pupal stage inside a capped cell. Immature bees whose cells have been capped with a brown wax cover by other worker bees; inside the sealed cell, the non-feeding larvae are isolated and can spin cocoons prior to pupating.

CAPPED HONEY: Cells full of ripe or mature honey, sealed or capped with beeswax.

CAPPED LARVA: The stage of brood development when the larvae turns into pre-pupae and is ready to have its cell capped and to spin its silk cocoon, about the tenth day from the laying of the egg.

CASTES: The three types of bee that comprise the adult population of a honey bee colony: the queen, workers and drones.

CHRONIC PARALYSIS VIRUS (CPV): Sometimes called chronic bee paralysis virus, this is a disease mainly of adult bees.

CELL CUP: The base of an artificial queen cell made of beeswax or plastic and used for rearing queen bees.

CLEARER BOARD: A board used to clear bees from a super prior to removing the frames ready for extraction.

CLIPPED QUEEN: Queen whose wing (or wings) has been clipped to stop her from swarming.

COLONY: A collection or family of bees living within a single social unit containing workers, drones and a queen.

COMB FOUNDATION: A thin sheet of beeswax with the base pattern of cells impressed on the sheet as a template for the bees to start making comb. Some foundation is also made of plastic.

COMB HONEY: Honey sold and eaten in the comb. Also referred to as cut comb honey.

DEFORMED WING VIRUS (DWV): A virus of the honey bee, often associated with a *Varroa* infestation, that causes newly emerged bees to have deformed wings.

DISEASE RESISTANCE: The ability of an organism to avoid a particular disease; primarily due to genetic immunity or avoidance behaviour.

DOUBLE: A hive that consists of two boxes.

DRIFT: The process in which bees join a different hive from their own, often due to loss of direction or when hives are placed too close together.

DRONE CONGREGATION AREA (DCA): An area where many drones from surrounding colonies gather to mate with queens during their nuptial or mating flights. The location of drone congregation areas remains stable over many years.

DYSENTERY: An abnormal condition of adult bees characterised by severe diarrhoea and usually caused by starvation, low-quality food, moist surroundings, or *Nosema apis* infection. (See also *Nosema*)

EMERGENCY QUEEN CELL: A queen cell in the middle of a brood frame that has been made quickly as a result of the queen bee dying suddenly.

ENTRANCE REDUCER: A device used to regulate the size of the bottom entrance. May be used to reduce hive entrance during winter months or during attacks from robber bees or wasps.

ESCAPE BOARD: A board having one or more one-way bee escapes in it. Used to empty a super of bees.

ESCORTS OR ESCORT BEES: Worker bees that are placed in a cage with a queen for shipping. Usually about four or five escort bees are included with the queen.

EUROPEAN FOUL BROOD (EFB): An infectious disease of honey bee brood caused by the bacterium *Streptococcus pluton*.

EXTRACTION: Removal of honey from comb. Typically refers to the use of an extractor but also includes non-mechanical methods such as crush and strain.

FAILING QUEEN: A queen that has reached the end of her fertile age and is not producing sufficient brood for the colony to survive. The colony is now ready for the supersedure of the queen.

FERAL BEES: Bees that are not kept in hives by beekeepers.

FERTILE QUEEN: A queen, inseminated instrumentally or mated with a drone, which can lay fertilised eggs.

GUARDING: Bees stationed at the entrance of a hive to detect and ward off invaders and to examine entering bees. Guard bees are typically about three weeks old.

HIVE STAND: A device that raises the bottom board off the ground and helps extend the life of the bottom board by keeping it dry.

HONEY BEE RACE: Honey bee race is a classification of honey bees, in particular the European honey bee, *Apis mellifera*, into different types. The most common races of *Apis mellifera* found in Australia are Italian bees, Caucasian bees, Carniolan bees, and dark German bees.

HONEY BOUND: A brood nest that is congested by cells/comb filled with honey, reducing the amount of space that the queen has to lay eggs.

HONEY FLOW: A time, usually in the spring and summer, when there are sufficient nectar-bearing plants blooming so that bees can store a surplus of honey.

HONEY SUPERS: The hive bodies or boxes, usually above the brood box, used by the workers to store honey.

HONEYDEW: The sweet secretion from aphids and scale insects. Since honeydew contains almost 90 per cent sugar; it is collected by bees and stored as honeydew honey.

HOOP PINE: A high quality pine wood native to New South Wales and Queensland and used to make supers and other hive parts.

HOPELESSLY QUEENLESS: A colony of bees that has been without a queen for several weeks and, as a result, is unable to make a new queen from young larvae.

HYDROGEN PEROXIDE: Found in all honeys and provides most of its antibacterial properties.

INBREEDING: Mating among related individuals that may cause genetic deformities in the bee.

INFERTILE: A bee that is unable to produce a fertilised egg, typically applied to a laying worker.

INTEGRATED PEST MANAGEMENT (IPM): A pest control method that uses a variety of complementary strategies including genetic, biological, cultural management, chemical management, as well as mechanical and physical devices such as screened bottom boards. IPM techniques are typically performed in three stages: prevention, observation, and intervention. It is an ecologically-friendly approach with the goal of significantly reducing or eliminating the use of pesticides while at the same time managing pest populations at an acceptable level.

INTERLOCKING: A design of wooden hive in which the corners of the boxes interlock with each other.

INTRODUCING: The process of introducing a new queen into an existing hive.

INTRODUCING CAGE: A small wood, wire or plastic cage used to ship queens and also sometimes to release them into the colony. Also known as a queen cage.

IRRADIATE: To irradiate a hive with gamma radiation to kill American Foul Brood or other diseases. There are few places across Australia that offer this. The facility is available in Melbourne, Sydney and Brisbane.

ISLE OF WIGHT DISEASE: *Acarapis woodi*, a small mite that causes tracheal infections of the honey bee, is believed to have been the cause of the severe decline in the honey bee population on the Isle of Wight, off the south coast of England, in 1904.

ISRAELI ACUTE PARALYSIS VIRUS (IAPV): First identified in Israel but now infecting honey bees in many parts of the world, one of several viruses that causes paralysis of adult honey bees.

ITALIAN BEE: A common race of bees, *Apis mellifera ligustica*, which originated in Italy. Bees have brown and yellow bands; usually gentle and productive, but tend to rob other hives. Italian bees are the most well known type of bee.

KASHMIR BEE VIRUS (KBV): A naturally occurring virus infecting *Apis cerana*, Kashmir Bee Virus jumped species to infect *Apis mellifera* and is now a common infection of this bee.

KILLER BEES: Africanized bee.

LARVA: The second stage of development in the lifecycle of the bee. The three stages are egg, larva and pupa. The larva stage is often called a grub.

LARVAE: The plural of larva.

LIFTING CLEAT: A wooden cleat attached to the side of supers and brood boxes to make them easier to lift.

LINE: A family or set of descendants bred from the same queen.

MANIPULATIONS: The process by which frames in a hive are moved or replaced.

MATED QUEEN: A queen that has gone on a mating flight or has been artificially inseminated with drone sperm.

MATING FLIGHT: A short flight taken by a virgin queen during which she mates in the air with several drones from other colonies. Queens usually mate with ten to twenty drones on one or more mating flights.

MIGRATORY BEEKEEPING: The practice of professional beekeepers and some hobby beekeepers to regularly move their hives to follow the flowering pattern of different types of trees.

MIGRATORY LID: The lid of a hive that has the same outside dimensions as the super. Called a migratory lid since it is of a convenient size to allow beekeepers to transport their hives on the back of trucks. A lid that does not extend over the sides of the hive.

MITICIDE: A chemical or biological agent which is applied to a colony to control parasitic mites such as *Varroa* or Tracheal mite.

NATIVE BEES: Bees that are native to Australia, for example, stingless bees. The European honey bee is not a native bee as it was introduced by Europeans in 1822.

NATIVE POLLINATORS: Apart from native bees, other animals may also be native pollinators such as birds or other insects. (See Native Bees)

NATURAL BEEKEEPING: A philosophy of keeping bees that puts the welfare of the colony ahead of the amount of honey collected during a season.

NATURAL HONEY: Pure honey that has not been heated or finely filtered during processing prior to bottling.

NECTAR FLOW: The time of the year when a tree or shrub produces a lot of nectar for bees to collect. The nectar flow of a particular tree often only lasts a few weeks, so professional beekeepers need to know what trees are flowering where and when and move their hives to follow the flowering trees.

NEONICOTINOIDES: A class of insecticides chemically related to nicotine.

NEST: A feral colony of bees that is not being managed by a beekeeper.

NEWSPAPER METHOD: A technique to join together two unrelated colonies by providing a temporary newspaper barrier between them. The bees will chew their way through the newspaper over a few days and the two colonies should merge together without fighting.

NOSEMA: A disease caused by protozoan spore-forming parasites living in the gut of adult bees. The two types of nosema that infect adult bees are *Nosema apis* and *Nosema cerana*. (See *Nosemosis*)

NOSEMOSIS: To be infected with *Nosema*.

NUCLEUS HIVE: A colony of bees housed within a small brood box usually containing only four or five frames. Nucleus hives are often used to rear queens.

NURSERY BEES: Young bees, three to ten days old, which feed and take care of developing larvae.

OBSERVATION HIVE: A hive with walls made largely of glass or clear plastic to allow the observation of bees at work. Often used as a teaching aid during courses.

ORGANIC HONEY: Honey that has been made from nectar from flowers grown in organically certified regions.

ORIENTATION FLIGHTS: Short orienting flights taken by young bees, usually by large numbers at the same time and during the warm part of day, in order for them to learn to fly and to familiarise themselves with their surroundings.

OVERCROWDING: A hive that has too many bees living in it for its size.

OVERWINTERING: The survival of a colony over winter.

OXYTETRACYCLINE (OTC): An antibiotic sold under the trade name Terramycin. Oxytetracycline is used to control American and European Foul Brood diseases.

PARENT COLONY: The home colony from which swarms or splits originated.

PARASITIC MITE SYNDROME (PMS): Signs of poor health in a colony of bees that has been infested with Varroa.

PHEROMONE: Several kinds of scents produced by bees to establish a basic form of communication or to stimulate a response.

PIPING: A series of sounds made by a queen, usually before she emerges from her cell.

PLANTATION PINE: The most readily available plantation timbers in Australia are softwoods such as hoop pine, radiata pine and slash pine, the latter two being introduced species.

PLASTIC FOUNDATION: Foundation placed inside a frame made entirely of plastic, used as an alternative to wax foundation. (See Plastic Frames)

PLASTIC FRAMES: Frames that are constructed entirely of plastic, including the outer rim.

PLAY FLIGHTS: Short flights taken in front and in the vicinity of the hive by young bees to acquaint themselves with the location of the hive. Play flights are sometimes mistaken for robbing or preparation for swarming. (See Orientation Flights)

POLLEN: The male reproductive cells of flowers. Pollen provides the protein in a young bee's diet and is frequently called bee bread when stored in cells in the colony. Pollen is an essential component of brood food. Honey is another essential component and provides the carbohydrate part of the bee's diet.

POLLEN BASKET: The area on the hind leg of bee adapted to carrying pellets of pollen or propolis back to the colony.

POLLEN SUBSTITUTE: A food material which is used as a substitute for naturally occurring pollen. Pollen substitute usually contains soy flour, brewers' yeast, powdered sugar, and other ingredients. Pollen substitute is used to stimulate brood rearing in periods of pollen shortage.

POLLEN TRAP: Device which forces the bees entering a hive to walk through a meshed screen, the pollen is brushed off the bees' legs by the screen and is collected from a collecting tray.

PRIME SWARM: The first swarm to leave the parent colony, usually with the old queen. (See also After-swarm)

PROPOLIS: A sticky resinous material collected from trees or other plants by bees; used to close holes and cover surfaces in the hive. Often referred to as bee glue. Propolis also has antimicrobial properties and is used by bees for this purpose.

PROPOLISE: To fill with propolis, or bee glue; used by bees to strengthen comb and seal cracks.

PROTEIN: Naturally occurring complex organic substances, such as pollen, and composed of amino acids. An essential food for brood to build body tissue before emerging as adult bees.

PUPA: The final stage in the life of a developing baby bee after larva and before maturity. Pupae are only found inside sealed brood comb. Pupae is the plural of pupa.

PYRETHRINS: A group of naturally occurring insecticides derived from the seed cases of the perennial plant Chrysanthemum cinerariaefolium.

QUEENLESS COLONY: A colony that does not have a queen.

QUEEN-RIGHT: A colony of bees that contains a laying queen.

RE-QUEEN: To introduce a new queen to a queenless hive. Usually to replace an old queen with a young one.

ROBBER BEES: Bees that rob other colonies of their honey. (See Robbing)

ROBBING: Bees that are stealing honey from other hives. This is a common problem particularly in autumn when nectar is not available in the field. The term also applies to bees that are cleaning out wet supers or cappings left uncovered by beekeepers.

ROPEY CHARACTERISTIC: A diagnostic test for American Foul Brood and European Foul Brood in which the decayed larvae or pupae form an elastic rope when drawn out with a matchstick.

ROYAL JELLY: Bee milk and worker jelly all refer to royal jelly, which is a pearly white, creamy substance produced by young worker bees to feed larvae. Royal jelly is a secretion of the hypopharyngeal glands located in the head of young worker honey bees.

SACBROOD: A viral disease of larvae, usually nonfatal to the colony, which interferes with the moulting process; the dead larva resembles a bag of fluid.

SCOUT BEES: Worker bees out searching for a new source of pollen, nectar, propolis, water, or a new home for a swarm of bees.

SCREENED BOTTOM BOARD: A framed screen used instead of a solid base in order to improve ventilation through the hive. Also used as a means to control *Varroa*, SHB and to allow debris to fall through to the outside of the hive.

SEALED BROOD: See Capped Brood.

SECTIONS: Small plastic or wooden boxes placed inside frames and used to produce comb honey.

SLUM-GUM: The black, sticky, fibrous, oily or waxy material left when comb has been melted and filtered by a beekeeper.

SMALL HIVE BEETLE (SHB): The Small Hive Beetle (*Aethina tumida*) is a destructive pest of honey bee colonies, causing damage to comb, stored honey and pollen. If infestation is sufficiently heavy, this may cause the bees to abandon their hive. The beetles can also be a pest of stored combs, and honey (in the comb) awaiting extraction. Beetle larvae tunnel through combs of honey, feeding and defecating, causing discoloration and fermentation.

SNOTTY BROOD: Sick brood with physical symptoms very similar to European Foul Brood (EFB). Snotty brood is also called 'Idiopathic Brood Disease Syndrome', which is a fancy way of saying that we haven't figured out what causes it yet and we are still looking.

SOLAR WAX MELTER: A glass-covered box in which wax combs and cappings are melted by the sun's rays and wax is recovered in cake form.

SPOTTED BROOD: An irregular brood pattern on the frame caused by disease or a failing queen.

SPLIT: To divide a colony in two to increase the number of hives.

SPRING BUILD-UP: The increase in the number of bees in a hive during very late winter and spring.

SPRING DWINDLING: A decrease in the size of the colony population during spring instead of the significant population growth which should be experienced. Spring dwindling is not associated with swarming in which about half the colony leave the hive to found a new colony.

STICKY: A super frame that has had its honey removed and is now covered in honey and is sticky to touch.

STREPTOCOCCUS PLUTON: Bacterium that causes European Foul Brood (EFB).

SUPER: A wooden box with frames containing foundation or drawn comb in which honey is to be produced. Named for its position above the brood nest, i.e. superstructure. The same size box is referred to as a hive body or brood box when it is situated below the honey supers and is to be used for brood rearing.

SUPERING: Placing additional supers on a hive in order to collect excess honey and provide more room for workers in a crowded hive. (See Oversupering.)

SUPERSEDURE: The replacement of a weak or old queen in a colony by a daughter queen. Shortly after the daughter queen begins to lay eggs, the mother queen disappears or is killed by the other bees.

SWARM: A collection of bees containing at least one queen that has left its home colony to look for a new nest. Swarms are currently without a home site and are looking for a new one, often resting on a tree or other object.

SWARM CELL: Queen cells usually found on the bottom of the combs before swarming.

SWARMING: The natural method by which bee colonies propagate. When the hive or nest becomes congested with bees in the spring, about half the colony flies off with the old queen to find a new home, leaving a new queen behind to head the old colony.

SWARMING SEASON: The time of year, usually between early spring and mid-summer, when colonies swarm.

SYRUP: A mixture of sugar and water used to feed bees when nectar sources are scarce or the beekeeper wants the colony to make honey for the winter.

TESTED QUEEN: A queen whose offspring have been tested to show that she has mated with a drone of her own race or has other qualities which would make her a good colony mother.

TOP FEEDER: A syrup feeder that is placed at the top of the hive under the lid.

TOP BAR HIVE (TBH): The top bar hive is a method to manage bees with removable combs which rely on top bars rather than frames for the combs. There are no standard dimensions for a TBH as there are for Langstroth hives. TBHs have some advantages for hobby beekeepers. (Also known as Kenyan top bar hive and Tanzanian top bar hive.)

TRACHEAL MITE: A small mite, *Acarapis woodi*, that infests the tracheas of the honey bee.

TRIPLE: A hive consisting of three boxes. This configuration may be one brood box and two supers, or a single brood box and two supers.

UNDERSUPERING: Placing an additional super beneath an existing super but above the brood chamber.

UNITE: Combining two or more colonies to form a larger colony.

UNRIPE: Honey in cells that have not been capped because the workers have not yet evaporated sufficient water to turn the nectar into honey. If unripe honey is extracted it can ferment.

UNSEALED BROOD: Brood during the egg and larval stages that are not yet in sealed cells.

UNSEALED HONEY: Honey in cells that have not yet been capped. (See also Unripe).

VARROA DESTRUCTOR: An external parasitic mite that attacks *Apis cerana* and *Apis mellifera* and develops in sealed brood cells. *Varroa destructor* is closely related to *Varroa jacobsoni*.

VARROA JACOBSONI: An external parasitic mite that attacks *Apis cerana* and *Apis mellifera* and develops in sealed brood cells. *Varroa jacobsoni* is closely related to *Varroa destructor*.

VARROA SENSITIVE HYGIENE (VSH): A genetically selected honey bee which appears to suppress *Varroa* mite reproduction.

VENT: A perforated piece of metal used to cover the air vent in the lid of a hive to stop robber bees or wasps entering.

VENTILATED LID: A hive lid that has vents in the rim allowing air to circulate.

VIRGIN QUEEN: A queen that has not yet left the colony to go on a mating flight and is thus not yet able to produce offspring.

WASHBOARDING: Worker honey bees exhibit a group activity known as rocking or washboarding on the internal and external surfaces of the hive. This behaviour is believed to be associated with general cleaning activities but under what circumstances workers washboard is not known.

WAX DIPPED: Hive parts that have been dipped in hot wax to protect the wood from the weather. Frames are not typically dipped in wax to protect them.

WAX GLANDS: Eight glands located on the underside of a bee's abdomen from which wax is secreted.

WAX MOTH: Usually refers to the Greater Wax Moth (GWM), *Galleria mellonella*, whose larvae bore through and destroy honeycomb as they eat out its impurities. Wax Moth infested hives are covered in a thick, spider-web like material. The Lesser Wax Moth (LWM), *Achroia grisella*, is the smaller of the two related species. Its larvae are not found in the large congregated numbers of the greater wax moth and are often solitary.

WEATHERTEX: A range of long lasting processed wood fibres that have been rolled into sheets. Weathertex is frequently used to make hive bases and hive lids.

WINTER CLUSTER: A closely packed colony of bees within a hive that forms a cluster to conserve heat when outside temperature falls below about 15° Celsius.

WINTER HARDINESS: The ability of some strains of honey bees to survive long winters by minimising their use of stored honey; the northern European black honey bee would be an example of this.

WINTERING DOWN: Preparing a hive for the winter.

WIRED FRAMES: Frames with wires holding sheets of foundation in place.

WORKER: An unfertilised female bee that makes up the majority of a colony's population.

WORKER JELLY: See Royal Jelly.

WSP: A super or brood box with a depth of 192mm. Named after the inventor William Stanley Pender (1866–1931), a well-known New South Wales beekeeper.

Acknowledgements and picture credits

A book such as this could only have been written with the assistance of many people. In particular I would like to thank my wife Barbara for the many hours she spent working on the book, offering suggestions, correcting my grammar and generally tolerating my absence from many domestic duties over the three years that it took to complete the book. Barbara also wrote Chapter 13, 'The bee-friendly garden'. My two daughters, Laura and Sian, spent many hours helping me take the many photographs that were required.

Special thanks go to Mary Ann, Daisy, Rosie, Lilly and Warren Higgs for letting me keep hives on their property, looking after them when I was busy with other work, taking many of the photographs in this book, assisting me in all manner of ways, and generally being very tolerant of my constant visits, which must have disrupted their lives considerably. Without the assistance and support of the Higgs family this book would not have been written.

Native bees are outside my scope of knowledge and I am therefore grateful to Erica Siegel AAPS, AFIAP, wildlife photographer (www.ozbirds-wildlife.com) for writing the detailed section on native solitary bees in Chapter 14. Some of the information on solitary bees was obtained from Dr Anne Dollin, Australian Native Bee Research Centre (www.aussiebee.com.au).

Identification of many of the native bees in the photographs was undertaken by Dr Michael Batley, Australian Museum, New South Wales, who also reviewed the chapters on native bees and other species of bees and wasps. Grateful thanks are also due to Dr Megan Halcroft, (www.beesbusiness.com.au), for generously allowing me to use material from the chapter she wrote on Australian stingless bees in the book *Pot-Honey: A legacy of stingless bees* (Springer, 2013). Dr Halcroft also reviewed the section on native social bees and offered many helpful suggestions.

Bron Woods of Bob's Beekeeping Supplies, Watsons Creek, Victoria, spent many hours reviewing the manuscript and imparted her considerable knowledge on beekeeping to improve the book's content. Cathrine Hall worked with me for many hours rearing queens and showed me how to use the Jenter method of queen rearing. Cathrine Hall also reviewed Chapter 1, 'The life of the honey bee'. David Briggs, queen rearer, Glenrowan, generously gave up his time to show me how he reared high-quality Italian and Caucasian queens commercially as well as reviewing the chapter on rearing queens. Jody Gerdts from Bee Scientifics (www.beescientifics.com) spent a lot of time discussing her experiences rearing *Varroa*-resistant queens while working at the University of Minnesota. Jody also reviewed Chapter 1, 'The life of the honey bee' as well as the chapter on rearing queens.

Thanks to Bob and Peter McDonald, Castlemaine, who lent me equipment to photograph as well as providing detailed reviews of the chapters on pests and diseases.

Ian Brown, who has 70 years' experience as a semi-commercial and hobby beekeeper, read the manuscript in its entirety and provided many valuable comments. Ian Brown was my mentor when I started beekeeping and is still the person I turn to when I need advice.

Dr James Makinson, Queen Mary College, University of London, spent long hours reviewing and providing information for the chapters 'The life of the honey bee' and 'Other species of bees and wasps'.

The chapters on the pests and diseases of the honey bee were the most exacting to write, particularly since many of the maladies, such as Tracheal Mites or *Tropilaelaps*, have not yet reached Australia. Much of the information for these chapters was obtained after meeting with Dr Eric Mussen, University of California, Davis; and Prof. Ernesto Guzman and Prof. Gard Otis, both of the University of Guelph, Canada. Dr Eric Mussen also gave invaluable assistance by providing a very detailed peer review of all of the pest and disease chapters. I would also like to thank Dr Mark Goodwin, Plant & Food Research, New Zealand, for taking the time to discuss with me the *Varroa* incursion into New Zealand and Dr John Roberts, CSIRO, who provided a detailed review of the pests and diseases chapters.

Martin O'Callaghan reviewed the chapter on sustainable beekeeping and allowed me to visit his home to take photographs of Warré and top-bar hives. Ron Rich reviewed the chapter 'Preparing honey for sale and competition'.

The second edition of this book was updated to include an extensive chapter on the Flow Hive. Preparation of this chapter would not have been possible without the generous assistance of Stuart Anderson, Martina Ryan and Bettina Clarke from Flow (www.honeyflow.com.au), who reviewed the text, patiently assisted when I had questions and provided photographs. I would also like to thank Fred Dunn, from the US, who read the draft and provided many photographs. My special thanks also go to Ben Moore of Ben's Bees, Victoria, who answered many questions, provided photographs and shared his extensive expertise.

Nadine Chapman, University of Sydney, provided recent research results on the origin and purity of Ligurian bees from Kangaroo Island as well as on *Apis mellifera mellifera* bees in Tasmania.

Thanks are also due to Prof. Ben Oldroyd, Bill Shay, Cindy Edwards, Howard Kirby, Anne Cliff, Roger Mitchell, and Steve Pernal, Agriculture and Agri-Food Canada, for providing information and/or reviews for different parts of the manuscript. George Bekier read the manuscript in its entirety and provided valuable comments.

Many other people also assisted in various ways with the preparation of this book — far too many to mention individually — but I would like to thank each of them collectively for their time and insights.

Finally an acknowledgement is due to the amazing bees mentioned throughout these pages. Without their influence this book would not have been written, and I hope the art and craft of beekeeping as described will assist readers in some small way towards an understanding their complex world.

PHOTOGRAPHS & DRAWINGS

Most of the photographs and drawings were taken or made by the author except for:

Adrian Iodice: 12.3, 12.15a, 12.15b

Alison Mellor: 14.23, 14.24

Barbara Owen: 13.1, 13.2, 13.5, 13.6, 13.8–13.11, 13.15, 13.16, 13.20, 13.21, 13.24, 13.32–13.42, 13.44–13.46, 19.19, 19.21

Ben Moore: 20.11, 20.33

Bronwen Harrington Geer: 2.22b

Chris Lahy: 5.4, 5.5

Chris Luck: 19.14, 19.15

Courtney Walshe: 10.6

CSIRO: 18.6–18.8, 19.9

Denis Anderson, CSIRO: 16.11, 16.13, 17.6, 19.1, 19.12, 19.23

David Briggs: 1.1, 1.6

Daisy Higgs: 3.15, 4.3–4.26, 4.31, 4.32, 5.12–5.17, 8.10–8.12

Dave Reede: 12.1

Dongdong Zheng: 4.33 to 4.36

Ernesto Guzman: 5.32, 15.3, 15.4, 15.15, 18.1

Erica Siegel: 13.15, 13.23, 13.24, 14.1, 14.2, 14.9–14.23, 14.25–14.27, 14.30–14.38, 14.40, 14.42–14.45, 14.47–14.49

Flow www.honeyflow.com.au: 20.1–20.5, 20.18, 20.22

Fred Dunn: 20.6–20.10, 20.12–20.17, 20.19–20.21, 20.23–20.32

Glenn Apiaries, USA: 19.3–19.5

George Vikentios: 13.43

Hamilton Gardens, New Zealand: 6.9

Hayley Carr: 6.6, 6.8

Ian Kimber: 17.1

Jason Critchley: 1.2, 1.15, 1.22, 1.27, 5.1, 5.2, 5.7, 5.28, 6.1, 6.13, 10.11, 10.12, 13.3, 13.6, 13.10–13.12, 13.16, 13.17, 13.25–13.28, 19.11

Joe Horner: 1.7, 1.8

John McLean, NZ: 1.14, 8.6

Julie Workman: 19.19

Kate Nolan: 1.9, 1.16, 1.17, 1.19, 1.21, 1.25, 1.26, 3.21, 5.25, 6.2, 7.2, 19.17

Ken Walker, Museum Victoria: 14.5, 17.16

Laura Matthews: 6.11

Laura Owen: 2.22a, 2.23-2.28

Lucy Wilks: 1.3, 1.23, 2.33, 9.9, 12.2, 13.4, 13.40, 15.16

Marie-Claire Desfontaines: back cover photo of author

Megan Halcroft: 14.4, 14.7

Marc Newman: 14.6

Nerina Lascelles: 3.16

Nick Monaghan: 14.39, 14.41

Biobees, New Zealand: 19.6

Oleksandr Holovachov: 13.20, 13.21

Paul Kozak, Government of Ontario, Canada: 15.8, 16.2, 16.4, 16.5, 16.8, 17.7–17.9

Patrik Berger: 14.24

Peter Boersch: 19.25

Paul van Westendorp: 15.11

R. Brito: 14.8

Russell Zabel: 14.3, 18.3, 18.4

Sian Owen: 2.29 – 2.32

Theresa Daley: 3.18, 5.3, 6.10

Trevor Monson: 7.1, 7.25

USDA: 1.5, 15.12, 15.13, 16.1, 16.3, 16.11

Vanessa Kwiatkowski: 6.7

Wikimedia — Sarefo: 18.5; Charles Lam: 19.2; Vera Buhl: 19.6; JJ Harison: 19.8; Tim Evison: 19.10; Frank Hornig: 19.13; Gideon Pisanty: 19.21; L. Shyamal: 19.24

Warwick Bone: 12.12, 12.13

Yeonsoon Bourke: 17.17

Zachary Huang: 5.33, 15.5, 15.7, 15.9, 15.10, 15.14, 16.12

Bibliography

GENERAL

Ben's Bees Facebook. https://www.facebook.com/BensBees/

Blackiston, H. 2015, *Beekeeping for Dummies*, Wiley, Hoboken, New Jersey, USA

Bonney, Richard F. 1990, *Hive Management*, Storey Publishing, Massachusetts, USA

Cramp, David. 2013, *A Practical Manual of Beekeeping*, Spring Hill, Oxford, UK

Cliff, Ann. 2010, *Bee Book: Beekeeping basics, harvesting honey, beeswax, candles and other bee business*, Aird Books, Melbourne, Australia

Flottum, Kim. 2011, *Better Beekeeping*, Quarry Books, Massachusetts, USA

Flottum, Kim. 2010, *The Backyard Beekeeper: An absolute beginner's guide to keeping bees in your yard and garden*, Quarry Books, Massachusetts, USA

Goodman, Russell and Kaczynski, Peter. 2015, *Australian Beekeeping Guide*, publication no. 14-098. Rural Industries & Research Development Corporation (RIDC), Canberra, Australia

Hughes, Craig. 2010, *Urban Beekeeping: A guide to keeping bees in the city*, The Good Life Press, Preston, UK

Johnson, Samantha and Johnson, Daniel. 2011, *The Beginner's Guide to Beekeeping*, Voyageur Press, Minneapolis, Minnesota, USA

Matheson, Andrew and Reid, Murray. 2011, *Practical Beekeeping in New Zealand*, Exisle Publishing, Auckland, New Zealand

NSW, Department of Industry, 2013, *Bee Agskills: A practical guide to farm skills*, NSW Department of Primary Industries, publication no. B121, NSW, Australia

Sammataro, Diana and Avitabile, Alphonse. 2011, *The Beekeepers' Handbook*, Comstock Publishing Associates, Ithaca, NY, USA

Warhurst, Peter. 2013, *The Bee Book*, self-published by Peter Warhurst, Queensland, Australia

Woodward, David. 2010, *Queen Bee: Biology, rearing and breeding*, Northern Bee Books, West Yorkshire, UK

BEE HEALTH

Aston, David and Bucknall, Sally. 2010, *Keeping Healthy Honey Bees*, Northern Bee Books, West Yorkshire, UK

Goodwin, Mark. 1999, *Elimination of American Foulbrood Disease Without the Use of Drugs*, National Beekeepers Association of New Zealand (NBANZ), Wellington, New Zealand

Goodwin, Mark and Taylor, Michelle. 2007, *Control of Varroa*, New Zealand Ministry of Agriculture and Forestry, Wellington, New Zealand

Pernal, Stephen and Clay, Heather. 2013, *Honey Bee Diseases and Pests*, Canadian Association of Professional Apiculturists (CAPA), Beaverlodge, Alberta, Canada

Plant Health Australia 2012, *Biosecurity Manual for the Honey Bee Industry*, Plant Health Australia and Rural Industries Research & Development Corporation (RIRDC), Canberra, Australia

Somerville, Doug. 2005, *Fat Bees, Skinny Bees: A manual on honey bee nutrition for bee keepers*, publication no. 05/054, Rural Industries Research & Development Corporation (RIRDC), Canberra, Australia

Somerville, Doug. 2014, *Healthy Bees: Managing pests, diseases and other disorders of the honey bee*, NSW Department of Primary Industries, NSW, Australia

LIGURIAN BEES ON KANGAROO ISLAND, SOUTH AUSTRALIA

Chapman, NC et al. 2016, Hybrid origins of Australian honeybees (*Apis mellifera*), *Apidologie*, 47: 26–34

Chapman NC. 2019, Genetic origins of honey bees (*Apis mellifera*) on Kangaroo Island and Norfolk Island (Australia) and the Kingdom of Tonga, *Apidologie*, 50: 28–39

SUSTAINABLE BEEKEEPING

Chandler, PJ. 2009, *The Barefoot Beekeeper*, Biobees, Milton Keynes, UK

Conrad, Ross. 2013, *Natural Beekeeping*, Chelsea Green Publishing, Vermont, USA

Heaf, David. 2010, *The Bee-friendly Beekeeper: A sustainable approach*, Northern Bee Books, West Yorkshire, UK

Magnum, Wyatt. 2012, *Top-Bar Hive Beekeeping: Wisdom and pleasure combined*, Stinging Drone Publications, Virginia, USA

Rowe, Tim. 2010, *The Rose Hive Method*, Green Hat Books, UK

THE BEE-FRIENDLY GARDEN

Blazey, Clive. 2012 *The Australian Vegetable Garden* (2nd Edition), New Holland Publishers (Australia) Pty Ltd, Sydney, Australia

Blazey, Clive. 2014, *There is No Excuse for Ugliness,* The Diggers Club, Dromana, Victoria, Australia

Hemphill, John and Rosemary. 2007, *Herbs: Their cultivation and usage*, Allen & Unwin, Sydney, Australia

Hooper, Ted and Taylor, Mike. 2006, *The Bee Friendly Garden*, Alphabet and Image Ltd, Devon, UK

Leech, Mark. 2012, *Bee Friendly: A planting guide for European honeybees and Australian native pollinators*, Rural Industries Research & Development Corporation (RIRDC), publication no. 12/014, Canberra, Australia

Little, Brenda. 2000, *Companion Planting in Australia,* Spring Hill, Oxford, UK

Little, Maureen. 2011, *The Bee Garden*, Spring Hill, Oxford, UK

Marshall, Tim. 2010, *Bug: The ultimate gardener's guide to organic pest control*, ABC Books, HarperCollins, Sydney, Australia

Rouble, John. 1999, *Backyard Composting*, Green Earth Books, UK

Urquhart, Paul. 1999. *The New Native Garden: Designing with Australian Plants*, New Holland Publishers (Australia) Pty Ltd, Sydney, Australia

Woodward Penny. 2012, *Pest Repellent Plants* (2nd edition), Hyland House, Carlton, Victoria, Australia

Wrigley, John and Fagg, Murray. 2002, *Starting Out With Natives,* New Holland Publishers (Australia) Pty Ltd, Sydney, Australia

Xerces Society. 2011, *Attracting Native Pollinators*, Storey Publishing, Massachusetts, USA

NATIVE BEES

Dollin, Anne, Batley, Michael, Robinson, Martyn and Faulkner, Brian. 2007, *Native Bees of the Sydney Region, Australian Native Bee Research Centre*, Richmond, NSW, Australia

Dollin, Anne. 2010, *Native Bees of Australia series*, Australian Native Bee Research Centre, Richmond, NSW, Australia

　　　Booklet 1 - Introduction To Australian Native Bees

　　　Booklet 2 - Nests of Australian Stingless Bees

Booklet 3 - Behaviour of Australian Stingless Bees

Booklet 4 - How to Recognise the Different Types of Australian Stingless Bees

Booklet 5 - Keeping Australian Stingless Bees in a log or box

Booklet 6 - Crop Pollination With Australian Stingless Bees

Booklet 7 - Tips on Stingless Beekeeping by Australian Beekeepers (Vol. 1)

Booklet 8 - Tips on Stingless Beekeeping by Australian Beekeepers (Vol. 2)

Booklet 9 - Boxing and Splitting Hives — A complete do it yourself guide for stingless bee keepers

Booklet 10 - Tips on Stingless Beekeeping by Australian Beekeepers (Vol. 3)

Klumpp, John. 2007, *Australian Stingless Beekeeping*, Earthling Enterprises, Queensland, Australia

NATURAL BEEKEEPING WEB RESOURCES

www.biobees.com

www.beekeepingnaturally.com.au

BEEKEEPING WEB FORUMS

www.beesource.com/forums

www.beemaster.com

NATIVE BEE WEB RESOURCES

www.aussiebee.com.au

www.beesbusiness.com.au

www.ozbirds-wildlife.com

www.sugarbag.net

www.zabel.com.au

FLOW HIVE

Flow: www.honeyflow.com.au, or www.facebook.com/flowhive/

The originators of the Flow Hive provide excellent information on their website both on the Flow Hive as well as on beekeeping. A good resource. Check out their YouTube videos as well.

Fred's Fine Fowl: www.fredsfinefowl.com

Fred, an experienced US beekeeper, has produced many excellent YouTube videos on Flow as well as beekeeping. Do a Google search for 'Frederick Dunn Flow Hive YouTube' and view his wide range of beekeeping videos.

Ben's Bees, Victoria, Australia: www.bensbees.com.au, or www.facebook.com/BensBees/

Ben writes a popular Facebook page on beekeeping and Flow Hives.

The Australian Flow Hive Beekeepers Group: www.facebook.com/groups/1381989935450304

A good source of information. I would recommend joining the group to read about the experience of Flow beekeepers and to ask any questions you may have.

In addition to the above, there is a lot of information on the Web. Do a Google search for 'Flow Hive'.

Index

Italicised page numbers indicate photos, diagrams, illustrations, or additional information in the captions.